IFCoLog Journal of Logics and their Applications

Volume 4, Number 11

December 2017

Disclaimer

Statements of fact and opinion in the articles in IfCoLog Journal of Logics and their Applications are those of the respective authors and contributors and not of the IfCoLog Journal of Logics and their Applications or of College Publications. Neither College Publications nor the IfCoLog Journal of Logics and their Applications make any representation, express or implied, in respect of the accuracy of the material in this journal and cannot accept any legal responsibility or liability for any errors or omissions that may be made. The reader should make his/her own evaluation as to the appropriateness or otherwise of any experimental technique described.

© Individual authors and College Publications 2017
All rights reserved.

ISBN 978-1-84890-246-6
ISSN (E) 2055-3714
ISSN (P) 2055-3706

College Publications
Scientific Director: Dov Gabbay
Managing Director: Jane Spurr

http://www.collegepublications.co.uk

Printed by Lightning Source, Milton Keynes, UK

All rights reserved. No part of this publication may be reproduced, stored in a retrieval system or transmitted in any form, or by any means, electronic, mechanical, photocopying, recording or otherwise without prior permission, in writing, from the publisher.

Editorial Board

Editors-in-Chief
Dov M. Gabbay and Jörg Siekmann

Marcello D'Agostino
Natasha Alechina
Sandra Alves
Arnon Avron
Jan Broersen
Martin Caminada
Balder ten Cate
Agata Ciabttoni
Robin Cooper
Luis Farinas del Cerro
Esther David
Didier Dubois
PM Dung
Amy Felty
David Fernandez Duque
Jan van Eijck

Melvin Fitting
Michael Gabbay
Murdoch Gabbay
Thomas F. Gordon
Wesley H. Holliday
Sara Kalvala
Shalom Lappin
Beishui Liao
David Makinson
George Metcalfe
Claudia Nalon
Valeria de Paiva
Jeff Paris
David Pearce
Brigitte Pientka
Elaine Pimentel

Henri Prade
David Pym
Ruy de Queiroz
Ram Ramanujam
Chrtian Retoré
Ulrike Sattler
Jörg Siekmann
Jane Spurr
Kaile Su
Leon van der Torre
Yde Venema
Rineke Verbrugge
Heinrich Wansing
Jef Wijsen
John Woods
Michael Wooldridge
Anna Zamansky

SCOPE AND SUBMISSIONS

This journal considers submission in all areas of pure and applied logic, including:

- pure logical systems
- proof theory
- constructive logic
- categorical logic
- modal and temporal logic
- model theory
- recursion theory
- type theory
- nominal theory
- nonclassical logics
- nonmonotonic logic
- numerical and uncertainty reasoning
- logic and AI
- foundations of logic programming
- belief revision
- systems of knowledge and belief
- logics and semantics of programming
- specification and verification
- agent theory
- databases

- dynamic logic
- quantum logic
- algebraic logic
- logic and cognition
- probabilistic logic
- logic and networks
- neuro-logical systems
- complexity
- argumentation theory
- logic and computation
- logic and language
- logic engineering
- knowledge-based systems
- automated reasoning
- knowledge representation
- logic in hardware and VLSI
- natural language
- concurrent computation
- planning

This journal will also consider papers on the application of logic in other subject areas: philosophy, cognitive science, physics etc. provided they have some formal content.

Submissions should be sent to Jane Spurr (jane.spurr@kcl.ac.uk) as a pdf file, preferably compiled in LaTeX using the IFCoLog class file.

CONTENTS

ARTICLES

Editor's Note 3557
    John Woods

Dale Jacquette: An Appreciation 3559
    John Woods

Identification of Identity 3571
    Jean-Yves Beziau

Inside Außersein 3583
    Filippo Casati and Graham Priest

Truth and Interpretation 3597
    Dagfinn Føllesdal

Content and Object in Brentano 3609
    Guillaume Fréchette

Nuclear and Extra-nuclear Properties 3629
    Nicholas Griffin

On the Existential Import of General Germs 3659
    Gudio Imaguire

How do we Know Things with Signs? A Model of Semiotic Intentionality 3683
Manuel Gustavo Isaac

Schopenhauer's Consoling View of Death 3705
Christopher Janaway

Taming the Existent Golden Mountain: the Nuclear Option 3719
Frederik Kroon

Reminiscences of Alonzo Church 3735
Nicholas Rescher

Noneism and Allism on the Objects of Thought 3739
Tom Schoonen and Franz Berto

Why Nothing Fails to Exist 3759
Peter Simons

Closing Words 3773
Tina Jacquette

Dale Jacquette. Publications 3775
Compiled by Tina Jacquette

Editorial Note

John Woods
University of British Columbia, Vancouver, Canada.
john.woods@ubc.ca

Dale Jacquette died on August 22, 2016, at the too-early age of 63. He was stricken by a pulmonary embolism at his home in Brügg and expired the following day. He leaves his wife Tina Jacquette (née Traar), his son Scott, and grandsons David and Jason. The present number of the IfCoLoG Journal of Logics and Their Applications has been assembled to honour Dale's memory and as a tribute to the substantial impact he has had on the philosophy of his day. In organizing the tribute, it was decided to give wide latitude to its authors. After all, not everyone who recognizes Dale's importance is a Jacquette scholar. My own view of the matter was that if we made the tribute solely an exercise in Jacquettian scholarship, we would lose the opportunity to expose him to an even wider philosophical audience. A further factor was Dale's extraordinary philosophical versatility. He was well-informed about virtually everything that has mattered in philosophy's long reign. In my last conversation with him the summer before his death, he had dazzling things to say about how Church's paradox might be recovered from, from which a further guideline would later emerge for the selection of authors. If someone of note had known Dale's personally, or had had professional contact with him, he would be free to write on anything that interested Dale. Other criteria were similarly flexible. If someone didn't know Dale personally or have professional contact with him, he would be welcome to contribute a paper on something he himself specializes in, provided that this too was an area in which Dale specialized. Equally welcome, needless to say, were papers that focused on some or other aspect of the Jacquette oeuvre. A final guideline was this. We would invite philosophers who, by acquaintance or reputation, Dale would be honoured to see in a publication such as this one.

I was first reader of the papers that appear here, and the task of second reader was mostly performed by one or other of our authors. There were a few occasions when an "outside" reader would be called upon. In that regard, my warm thanks to Johan Marek, Mark Migotti and Daniel Parrochia.

I am grateful to general editor Dov Gabbay for the hospitality of his journal and to Jane Spurr for her assistance in readying this number for publication. I am especially indebted to Tina Jacquette for her support of the project, and for her timely help on details of Dale's early academic life. My largest and proudest gratitude is to the talented philosophers who have made this number the philosophical success that we had all set out to make.

Dale Jacquette: An Appreciation

John Woods
University of British Columbia, Vancouver, Canada.
john.woods@ubc.ca

Dale Jacquette was Ordentlicher Professor mit Schwerpunkt Theoretische at Universität Bern (Senior Professorial Chair in Logic and Theoretical Philosophy). Born in Sheboygan, Wisconsin, he was educated there until high school matriculation and admission to Oberlin College. He graduated from Oberlin College in 1975 with an Honours BA in Philosophy, under the supervision of Robert Grimm. His honours thesis was entitled "Aristotle on identity and individuation". That same year he was elected to the prestigous Honour Society Phi Beta Kappa. Instead of taking a "gap year", which was then coming into vogue, he spent a year at Temple University talking to Bill Wisdom and Hugues Leblanc, and followed with a year at Leeds chatting with Peter Geach, in each case enlarging his appreciation of formal methods in philosophy. He then enrolled in the philosophy programme at Brown University and earned an MA in 1981 under the supervision of Roderick Chisholm. Two years later he received his PhD, also directed by Chisholm, with a dissertation entitled *The Object Theory Logic of Intention*. Before his Bern appointment in 2008, he held positions at Pennsylvania State University, University of Nebraska at Lincoln, and Franklin and Marshall College.

At his death, Dale Jacquette was the author, editor or co-editor of 34 books, and left four research monographs and one edited volume unpublished. Of particular note to me is the massive *Frege: A Philosophical Biography*, now in production with the Cambridge University Press. I say "of particular note" because I was one of the book's external readers, and was rewarded with a much richer and nuanced understanding of Frege than I had previously possessed. The Frege book displays Jacquette's talents and interests on a wide canvas — a tenacious grasp of historical context, a sharp eye for expositional and biographical detail, a good grasp of German, a well-informed acquaintance with late 19th century mathematics, and with Frege's anxieties, both mathematical and philosophical, about the foundational security of arithmetic. Also evident in this manuscript, as in Jacquette's other writings, is a willingness to develop and hold views that press against received opinion. This is a typical Jacquettian virtue. In his voluminous writings, he is a "follow your nose without fear or favour" kind of philosopher, and little inclined to be part of the in-crowd, except when he agrees with it. Judging from his published books alone, his range was astonishing, especially in a world in which one-trick pony scholarship is both welcome and, one might even think, rather preferred.

Also in posthumous production is Jacquette's edited *Bloomsbury Companion to the Philosophy of Consciousness*, in the Introduction to which he quotes Kingsley Amis' "Consciousness was upon him before he could get out of the way." It is a wise editor who would say, "There is thankfully no party line philosophically in consciousness studies." Perfect! This is especially evident in the volume's attractive mix of authors, several of whom first attained their philosophical reputations in areas not especially, or at all, connected to the puzzles of consciousness. In some areas of scholarship, the same handful of authors, representing such rival points of view as chance to be, show up repeatedly in one handbook, collection or conference proceedings after another. In a way, these are closed-shop enterprises, displaying more in the way of business as usual than innovative derring-do. The philosophy of consciousness is different. It is a gold rush, in which new paradigms are launched and original claims are staked. That's what we see amply in view in this welcome Companion.

Jacquette's versatility and industry are abundantly in play in 210 published articles and six others he left unpublished at his death, one of which has now appeared in *Brentano Studien*. Also impressive are the 119 contributions to books, and the 13 further ones not yet published. Here, too, the range is striking — Condillac, Husserl, Kripke, Humboldt, Socrates, Turing, Plato, Newton, Tarski, Mally, computability, *petitio principii*, category systems, Collingwood, Searle, Anselm, violence, diagonalization, paradoxes, cannabis, Burleigh, Reid, Kant, fiction, agenda relevance, quantum indeterminacy, paraconsistency, Aristotle, Goodman, Flaubert, Descartes, Twardowski, Borges, Bosaquet, analogy, Barthes, the square of opposition, socioeconomic Darwinism; and more. It is frequently said that there was a time when a learned and determined scholar could know everything that mattered, and that it is only since the explosions of the Enlightenment that this kind of learnedness hasn't been considered remotely conceivable. Even the most widely learned of our own day might have a hard time imagining that a large learnedness is still not only conceivable but also at times convincingly realized. Perhaps the nearest versatility-comparison in contemporary English-speaking philosophy is the remarkable body of work by Nicholas Rescher.

One of the standing biases against wide-scope scholarship in philosophy arises from an understandable worry about dilettantism, attended by the suspicion that anyone who writes about subjects as diverse as Meinong, Brentano, Wittgenstein, Berkeley, Hume, Boole, Russell, Frege, Schopenhauer, Tarski, Rescher, the philosophy of mind, ontology, journalistic ethics, capital punishment, possible worlds, the philosophies of logic, mathematics, religion and knowledge, isn't likely to be much good at any of them. There is, however, plenty of evidence that people who actually make the effort to publish the results of their wide-spread interests often stand as exceptions to these generalizations. Jacquette was never unaware of the bias against widespreadness, and he wrote with a depth and a confident determination to show that his own work was no confirmation of it. He was indeed a contrary-minded philosopher when he wanted to be, and it reflects well on him that he was an "eyes wide-open" one.

In 2007 Jacquette did something dangerous. He published a new translation and critical commentary of Frege's *The Foundations of Arithmetic*. In so doing, he put himself in competition with the much loved Austin translation of 1950. I myself don't think that the rivalry was intended, except possibly in a double-effect sort of way. Jacquette wasn't out to show up Austin, still less to take him down. Rather he wanted to relieve an itch of his own. Jacquette thought he had a good translation in him, and he wanted to bring it to the surface. One of the critics took issue with how the new translation handled certain passages of Frege. He ended his review by announcing that he himself would stay with Austin's. In a way, that wouldn't have bothered Jacquette. His objective was to produce a translation which he himself thought well of, and he was not interested in convincing people to stop reading the other one. Think here of the *Tractatus*, whose first English translation was Ogden's (although he seems it was at least as much Ramsey's). Sixty-three years later came the Pears and McGuiness. Some commentators think the Pears-McGuinness is more faithful to the German, and others think that the Ramsey-Ogden better captures the music of Wittgenstein's meaning. The objective of Pears and McGuinness was not to send the predecessor-translation into retirement. It was to make available a further instrument for getting to the bottom of Wittgenstein. Both translations are widely used to this day. That too was what Jacquette was hoping for. It is significant that in his entry in *The International Directory of Logicians* (2009), of the twenty-four listed as main publications, Jacquette gives pride of place to the Frege translation. I see in this an admirable and durable self-possession.

We all know philosophers who have more ideas than you can shake a stick at. They attend departmental colloquia on any and all topics, and they invariably ask the best questions and make the most helpful suggestions, but rarely write up these insights for publication. Richard Cartwright — also an Oberlin-Brown alumnus — was like this. He was, hands down, my most stimulating and widely ranging doctoral teacher, but his publication record was comparatively slight, especially in relation to the very high regard in which he was held. Cartwright was a "make haste slowly" kind of philosopher, He thought that philosophy's natural home was the common room, or conference table, where its transactions would be recorded in the memories of those present at the time.

Other philosophers are more like artists. A painter who speaks wonderfully about his painterly thinking is no painter until he paints. A composer who never gets around to writing the score of the music that runs through his head is its composer in name only. A painter who won't show his work is a painter manqué, and a composer who declines to have his music played is an artist who's selling real estate. Philosophers are sometimes like this. They won't see themselves as philosophers unless they put their thinking on the record, that is, into cold or digital print. If such a philosopher is as versatile as Jacquette was, it only stands to reason that he will publish a lot and far-rangingly. In the present-day academy, notwithstanding its massive publish or perish pressures, there are more Cartwrights than Jacquettes. In intellectually robust communities, there is plenty of room for both. It would do everyone some good if there were a greater balance between the philosopher as performing artist and the philosopher as no-hurry tactician.

Jacquette was first and foremost a logician, and in the *Directory* entry his dissidence is expressly acknowledged. He writes that the purpose of his intensional approach to logic was to provide a more general intuitive semantics for propositions about nonexistent entities than was available in standard extensionalist logics and semantics, ensuing from Frege's *Grundgesetze*, Whitehead's and Russell's *Principia*, and Tarski's advances in the theory of models.

Regarding his extensive writings onlogical and semantic puzzles, including the less discussed ones such as the paradoxes of Grelling and Pseudo-Scotus, and his own variation, "the soundness paradox", Jacquette writes that his "policy toward them has been most profoundly influenced by Wittgenstein in both his early and later periods ... [and he] expects logic to cure itself as Wittgenstein had maintained in the *Tractatus Logico-Philosophicus* and *Notebooks* 1914-1916. This is the limited sense, especially in confronting the paradoxes, in which [my] approach to logic remains Wittgensteinian". At our last lunch , in Istanbul in June of 2015, we also chatted briefly about Franz Berto's reflections on Wittgenstein's reservations about the Gödel sentence is paradoxical. Dale applauded Franz's balanced treatment of the question, and added that he himself would have been more sympathetic to Wittgenstein's. We also found ourselves agreeing that, whatever its *bona fides* in the 1931 proof, there is no truth of Peano arithmetic that the Gödel sentence represents. It may well be a true sentence of *formalized* arithmetic, but not a true sentence of *number theory*. In his vision statement Dale adds:

> Intensional logics in particular will increasingly take precedence over traditional extensional systems.

Any reader in 2018 familiar with the development of logic since *Principia* and

C. I. Lewis' axiomatizations of modal propositional systems will know, this to have been an understatement. In another entry in the *Directory*, Timothy Smiley writes,

> Under the umbrella of 'mathematical logic' I see a mass of mathematically-driven work without even a vestigial connection with the theory of argumentation. When the caravan has moved on, I hope logic will be left to return to its roots.

In Istanbul, Jacquette reminded me of this and saluted Smiley for having said it. I thought at the time that the salute more courteous than heartfelt. Jacquette never waivered in his trust of formal semantics to do the heavy philosophical lifting in the domains for which it is suitably contrived. But he was no come-what-may mathematical loyalist. Greatly to Jacquette's credit is the tenacity with which he probed inadequacies in the mathematics of formal logics. Especially important, and technically adroit, are his reservations about the role of diagonalization in logic and mathematics, and the inadequacies he finds in Tarski's truth-schemata, for example, in his 2010 paper in the *Journal of Logic, Language and Information*.

Although he leaves a legacy, Dale Jacquette was never, I think, a legacy-intending philosopher. I think that his most enduring influence will be his Austrian writings. I mean by this his work on the approaches to logic and ontology developed in the period from Brentano, Meinong, Husserl, and Mally. Husserl, of course, was German, but thanks to the instruction he received from Brentano in Vienna as a mathematician attending the lectures of a philosopher, he counts as an honourary Austrian. Austrian phenomenology had a distinguished earlier presence in pre-Frege Jena and a rival one in Frankfurt am Main, but one of Husserl's signature achievements was his recognition of psychology's importance for logic, not excluding the logic of mathematics, a view Dale vigorously endorsed in his edited volume *Philosophy, Psychology and Psychologism*, which also appeared in 2010.

At the heart of the Austrian movement was the insistence that thinking in the absence of aboutness is nothing. Alongside comes the implication that extensional logic, in the manner of Frege's second-order functional calculus for the foundations of mathematics, was untrue to the facts of lived reasoning experience. It fails to capture the ways in which human beings think, reason and argue. It fails on purpose. It was never Frege's intention to produce a logic of how people think, echoing a remark of Peirce about logic in its Òstrict" sense. It prompted Quine's admiring observation that, while logic is an ancient subject, it's been a great one only since 1879. We could say that the semantic heart of Jacquette's intensional logic is the plain and obvious truth that there are lots of things don't exist. When the suggestion was put to Quine, he pretended not to understand it. Later I came to see that Quine had no use for the being-existence distinction, thereby rendering moot the question of whether reference and quantification are existentially loaded. For Jacquette, on the other hand, the distinction is vital. Not only does he want to preserve the commonplace fact that there are plenty of things that don't exist but, for more theoretical reasons, he doesn't want to deny himself the means at least to *formulate* the still unsolved problem of the non-existent existing golden mountain.

If we paid intellectually conscientious attention to what is actually happening when human beings have thoughts, we'd not fail to notice the frequency and effortlessness with which we know that what we're thinking of doesn't actually exist. This strongly suggests that when we stand in the aboutness relation to something, we bear a real relation to an object which might be an impalpable one. Consider here Jacquette's reference to the creatures of fiction. Whatever our other intimacies with him might turn out to be, one of the things we can't do with Sherlock Holmes is have him to tea or he with us. More generally, when we invoke something fictional, x say, we might not invoke something real. But no one need doubt the reality of the relation and the impalpability of its referent. Fiction is an especially interesting case. It seems to make possible real connections between reality and unreality. What is sometimes overlooked is what brings such contacts about. Again, if we carefully attended to empirically discernible facts of fiction-making, we see that the *world* is the maker of the unreality that fiction is. Writing at his desk in Britain, it was *Doyle's* doings which brought about the Holmes stories, thereby bringing about the objects and events that the stories relate. In so doing, Doyle made no addition to his country's population and produced no new entries for the history of English criminal detection. Doyle stands to Holmes in the relation of having created an unreal man. This is arguably the most foundational of the real relations which the real sometimes bear to the not real. It turns out, however, that this was not Jacquette's view of the matter.

Jacquette's is a neo-Meinongean logic. Meinongianism without the Òneo-"has an unrestricted comprehension axiom. Its more qualified version restricts the comprehension principle to one which respects Mally's distinction between an object's nuclear and extranuclear properties, which remains one of the most fraught distinctions in the Meinongean repertoire. Even so, one way or the other, the universe of Jacquette's logic remains enormous. It is everything there is. But isn't this what Quine said? Yes, but he said it for the purpose of the preferred objects of his ontological commitments which, in the end, would be all and only those entities needed for the truths of mathematics and the most mature precincts of the physical science, subject only to the canonical requirement of classical first-order reformulation.[1] Jacquette, on the other hand, suffered from none of Quine's ontic-

---

[1] As it turns out, quantum physics wouldnÕt play extensional ball, something that Putnam was

anxieties, and he had no taste for desert landscapes in metaphysics or logic. He had instead a large and welcoming appetite for ontological abundance. His is a domain of such encompassing inclusivity as to raise in some quarters suggestions of promiscuity. What is needed therefore is a formal apparatus for keeping a disciplined check on how what interacts with what, and in what inference relations they are eligible for engagement. *Meinongian Logic* (1996) develops this technical machinery with precision and rigour. It is a more technically realized book than Terence Parson's technically groundbreaking one sixteen years before. In both these works, neo-Meinongian logic provides a general-purpose semantics for everything whatever that doesn't exist, or didn't or won't or couldn't. It is a virtue of the approach that nonexistence is afforded a wholly general and unified treatment with which to avoid *ad hoc* measures. The same rationale pervades the several other versions of noneism in the descendent class of Routley and Meyer, and later Priest and Berto. Here I find myself oppositely inclined. Fiction has peculiarities that nothing else in a Meinongian ontology has, in a way that puts at risk the adaptation of a theory of all non-existent ones to an account of the fictional ones.

One feature of neo-Meinongianism that has attracted some plausible criticism is that all objects of whatever ontic status pre-exist any subsequent creative activity here at home. It is true at home that Vulcan was hypothesized to exist by Le Verrier (actually by Babinet), but for Jacquette, it is not true that Vulcan is a hypothetical object. Vulcan was an object constituted by pre-existing properties. Vulcan was "there" long before Le Verrier and Babinet drew their first abductive breaths. In a Jacquettian semantics, it is the same way with fiction. It provides that, while Sherlock was fictionalized by Doyle, he was not the object of Doyle's own creation. He was the object of his constituting properties, not one of which was created by Doyle. These implications strike some commentators as odd. I myself think that they are more troublesome than odd. They reveal two competing impulses in Jacquette's intensional logic. One is its attachment to Brentano's and Husserl's psychologism, in which real-life cognitive experience plays a central role. Another is the platonistic pull of Meinong's theory of objects, in which human thinking has no inherent role to play. The two strands don't easily or coherently braid.

One of the breakthrough achievements in late-19th and 20th century semantics was the launching and perfecting of the mechanisms that drive the model theories of purely artificial languages. Misleadingly dubbed "formal *semantics*" by Tarski, model-theoretic approaches have long since been *de rigeur* in exact philosophy. Dummett has something important to say about this. Having noted the necessity of mathematical tools in structuring a system's model theory, he regrets

---

onto as early as 1968.

the concomitant loss of focus on what model theory was wanted for in the first place. In Jacquette's approach, model theory is wanted for the systematization and precisification of Brentano's and Husserl's insights into human experiencings of aboutness. Therein lies a tension. At what point should we allow the mathematics of model theory to override empirically discernible facts of our aboutness-experiencings and the truth-making conditions under which these aboutness relations depend? Scientists know a like tension between what their subject-matters ask for and what their theoretical mechanism provide for them. Dummett is of the view, and Smiley too, that the present-day trend is for technically proficient philosophers to prefer tools to subject-matter. I mention this here to flag a more general methodological point. Model-theoretic semantics are formal representations of properties of natural language that have caught our philosophical eye. Formal representations always distort the properties they represent, variously so. The risk they carry is the potential to make these target properties unrecognizable in formally representing the model theory. It is a standing liability and a general one, and in no way peculiar to Jacquette's intensionalist semantics. So it would be neither unfair nor unfriendly to ask whether there is any significant degree to which this same assessment might hold for Jacquette's logic of nonexistence? Yes or no, the fact remains that *Meinongean Logic* is a masterly achievement and a notable effort to evade Dummett's worry.

I come now to the question of how Dale Jacquette came to be the formidable thinker he was. There is little doubt of Chisholm's influence. When Jacquette was his student in Providence, analytic philosophers had plighted their troth to the decomposition of concepts, which would render them into their simple and unanalyzable components in ways that would give us a greater *à priori* understanding of them. The results of such analyses would be conceptually necessary truths, and in some quarters of enquiry, would be enshrined as axioms, both insusceptible of proof and immune from overthrow. In the preface of Russell's *Principles of Mathematics* (1903), published the year following his announcement to Frege of the paradox that destroyed the conceptual truth embodied in *Grundgesetze* I's Basic Law V, Russell declared that without Moore's having led him from idealism to what came next, it would have been impossible for Russell to say anything philosophically coherent about the foundations of arithmetic. But he said this in the very aftermath of the death of the analysis that gave us the indemonstrable and irrefutable Basic Law V. Russell was never thereafter (how could he be?) an analytic philosopher in Moore's sense.

In the United States, Moorean analysis never held centre stage, notwithstanding the attention lavished on Frege, Russell and Moore. Pragmatism was up and running at Harvard and Johns Hopkins, and critical realism, spurred on by Wilfrid Sellars' father Roy Wood, was flourishing at Michigan. When Henry Johnstone Jr., a much respected colleague of Jacquette's, arrived at Penn State, he had come from Middlebury College, whose philosophy department had been more than a little sympathetic to idealism. Many years ago my teacher and colleague David Savan, who had been one of Quine's students at Harvard, told me that he thought that Quine, too, was an idealist. When I was a Ph.D. student at Michigan, in a seminar on Morton White's *Toward Reunion in Philosophy*, William Frankena warmly welcomed this idea. Meanwhile, logical positivism had arrived from Europe and attained a considerable foothold for a time. But there is little that's recognizable as conceptual analysis in the writings of Carnap, Hempel, Reichenbach and the others, but plenty of evidence of the "rational reconstruction" of philosophically evasive concepts. Major developments were stirring in places such as Pittsburgh, where Sellars *fils* was doing his best to destroy the myth of the given. Fruitful coalitions emerged with the pragmatism of Harvard, the inferentialism of Hilbert and Brouwer and the idealism of Hegel. Rescher is a pragmatic idealist, and Macdowell and Brandom were socially pragmatic inferentialists. However, for the most part, these non-analytic developments had no deep anchorage in Austrian phenomenology.

At Brown, when Chisholm published *Perceiving* in 1957, there was little evidence of a systematized phenomenology. 1976 was a bit different. *Person and Object* appeared just as Jacquette was in Philadelphia and Leeds preparing himself for Brown. In 1981, in the sweet part of Jacquette's time in Providence, Chisholm issued *The First Person: An Essay on Reference and Intentionality*. We may therefore take it that Chisholm's influence on Jacquette is unmistakable. Less clear is where Jacquette's formidable model-theoretic skills arose, I mean took hold in a deeply operational way. Certainly he will have known of Chisholm's respect for formal methods in philosophy, but Jacquette's formal methods are nowhere discernible in Chisholm's work. Jacquette had a large and friendly respect for Henry Johnstone and Salim Kermal at Penn State and, over the years for Rescher, some 138 miles due west of University Park. It was Rescher who recommended him as his editorial successor at *American Philosophical Quarterly* and who was helpful in Jacquette's relocation to Bern. Jacquette read widely and profitably, and had a good eye for important work. He was quick to acknowledge his debts and generous in doing so. But in the famous words of a former Deputy-Prime Minister of Canada, he was Òno one's baby". He admired the philosophers he learned from but had no inclination to immerse himself in their intellectual *personae*. He admired Chisholm but was not a Chisholmian. The same is true of Rescher and the several others. Rescher is an idealist influenced by Kant and Leibniz. Jacquette was drawn to Austrian realism, and thence to Frege and Wittgenstein. The only "ianism" he ever exemplified was Meinong's. Even there, he was his own kind of Meinongian. In *Meinongean Logic*, there are two footnotes to Rescher and none to Chisholm. In

Alexius Meinong, Chisholm merits 16 mentions and Rescher one. Later Jacquette developed an agreeable acquaintance with Dagfinn Follesdal, in connection with their involvements with the Lauener Foundation. Follesdal is the modal logician who was instrumental in bringing Husserl to the attention of philosophers in Palo Alto, Oslo and elsewhere. But well before he moved to Bern, Jacquette had a fully developed intellectual character of his own. Indeed Follesdal makes no appearance in the index of *Meinongian Logic*, and makes no like appearance those nineteen years later in *Alexius Meinong*. The sole reference to Brandom is to his co-authorship with Rescher of their 1980 book on paraconsistent logic in 1980. I infer from this that from early on Jacquette was running an operating manual mainly of his own crafting.

In the forthcoming book on Frege, Jacquette reflects on the influences that brought Frege to an intellectual maturity that would trigger a revolution in philosophy. Only twice in Frege's time at school and university did he have a course on philosophy or logic. The teacher of one had written on logic but was mainly an aesthetician. The other offered a run of the mill course in what we now see as pre-Fregean logic. Frege was a mathematician who shared, but did not originate, worries about the foundational stability of mathematics. He had satisfied himself that all of mathematics could be grounded in arithmetic, provided that number theory could supply its own foundations. Thinking that it couldn't, Frege concluded that either arithmetic is groundless or its foundations lie elsewhere. He knew just enough about the boilerplate logic of the day to realize that it couldn't do the job for arithmetic. So, to put it simply, he was left with no option but to think up a logic that would serve this purpose, the second-order functional calculus. Jacquette leaves the inference that Frege was mainly a thinker of his own foundational making. That would indeed make Frege the *pure laine* prodigy I too think he was. In my post-mortem reflections on our sadly departed friend, I've been drawn to a like view of him. Although not the originalist that Frege was, Dale Jacquette was an independent man of his own intellectual making, and in the end, he was more an Austrian philosopher than an American one.

For all his sheer busyness, Jacquette was a man of parts. He had an easy and welcoming nature, and was an accomplished photographer and an informed lover of music and the visual arts. He was hooked on travelling. He would go anywhere at the drop of a hat, and he favoured doing it the hard way whenever possible — by hiking, biking and always opting for deep-steerage economy class when he flew. His magnificent lectures — as witness his 2015 talks on Boole in Istanbul — were models of organization and intellectual clarity. He greatly enjoyed the discussions that followed. But he was cautious about his intimacies, none more heartfelt, nourishing and enduring than his love for his wife.

Dale Jacquette: An Appreciation

For assistance in preparing these remarks, I warmly thank Nicholas Rescher, Manuel Gustavo Isaac, Paul Bartha, Hilary Gaskin, Guillaume Fréchette, Fred Kroon and Tina Jacquette. This essay is a revised and extended version of my obituary of Dale, which appeared in *Brentano Studien* in 2016. It appears here with permission.

Received 11 December 2017

Identification of Identity

Jean-Yves Beziau
University of Brazil, Rio de Janeiro, Ecole Normale Supérieure, Paris
jyb@ufrj.br

1 Identity: between contradiction and tautology

Wittgenstein wrote in the *Tractatus* (5.5303): "to say of *two* things that they are identical is nonsense, and to say of *one* thing that it is identical with itself is to say nothing at all". From this point of view, we can say that identity is swinging between the Charybdis of contradiction and the Scylla of tautology. Identity is a fundamental process in thought. Though is moving in a subtle way and it is important to understand how it works, how it finds its way between Charybdis and Scylla.

When we are thinking, we are identifying things which are different. Identity and difference go together. A *concept*, like for example the cat concept, consists in identifying a series of different objects. Among animals, we are unifying a class of them who have the same features or properties. At the same time, we know that each cat is different from the other ones. We are making abstraction of these differences. For *individuality* there is also an identity process. When we are talking about the Amazon river, we are identifying by abstraction a series of phenomena out of time and space. To understand how these two identification processes work — conceptualization and individualization — is to understand the Charybdis of identity.

The Scylla of identity can be examined through what we can call trivial identity which is described by the following version of the principle of identity: *everything is identical to itself and different from other things*. Sometimes the principle of identity is expressed only as the first half of this sentence: *everything is identical to itself*. Surely this is not enough to characterize identity since any reflexive relation obeys this axiom. Identity has to be reflexive but *only* reflexive. The second part of the principle of identity does not help to solve the problem, because *different from other things*, means different from different things, or not identical to non identical things, it is a tautology. At the end *everything is identical to itself and different from other things* is equivalent to *everything is identical to itself*, that is to say just reflexivity, which is not enough.

In a sense, trivial identity is inexpressible in ordinary language and it is also non axiomatizable in first-order logic (see e.g. Hodges 1983), but at the same time it can be expressed through mathematical diagrams: the diagonal of the Cartesian product of a set or with circular arrows around each point of a graph. The Scylla of identity can therefore be visualized. It is quite nothing, but it is something.

2 Naming identification and naming differentiation for singular terms

Quine wrote about the above quote of Wittgenstein that the logico-philosopher was mistaken and that "Actually of course the statements of identity that are true and not idle consist of unlike singular terms that refer to the same thing" (*Word and Object*, p.117). So the question of reference would be a way to navigate through the Scylla and the Charybdis of identity. But Quine is touching here only one aspect of the problem, there are many other ways to tame the two monsters. The same thing can be named by different names, but also different things can be named by the same name, and these two processes do not concern only singular terms, but also plural terms.

Brasilia and *the Capital of Brazil* can be considered as two different names for the same thing: the city of Brasilia. *Rio de Janeiro* can be considered as the same name for two different things: the city of Rio de Janeiro as it is now and the city as it was before. We can call the first process *naming differentiation* and the second one, *naming identification*.

Now we have to understand why we are doing so. This use of language certainly reflects some fundamental processes of thought that have to be clarified. Frege wanted to avoid this use of language, to develop a perfect language where there will be no such ambiguities (Frege 1883), and Wittgenstein was following the same line when writing: "Identity of the object I express by identity of the sign and not by means of a sign of identity. Difference of the objects by difference of the signs." (*Tractatus*, 5.53).

But in fact these ambiguities are part of the very nature of thought, we don't have to get rid of them — it is too easy, a kind of bulldozer methodology — they have to be properly understood. This is in some sense what people have tried to do in the stream of philosophy of language, using in particular the distinctions between sense and reference, definite descriptions and proper names, and the notion of possible world. This may give interesting hints but does not explain everything.

The Capital of Brazil was a name for *Rio de Janeiro* in the state of the world in 1950 and is now in the present world a name for *Brasilia*. We are performing a naming identification, using the same name for two different things. This is based on a clear relation between two possible worlds, the time line, and on the fact that to be a capital of a country is a precise feature that does not characterize a city and may vary through time.

The Capital of Brazil is considered as a definite description and one may think that it is its very nature of definite description that explains the naming identification. By opposition, some people, in particular Kripke, have argued that a proper name like *Rio de Janeiro* would always refer to the same thing in all variations of the world, in all possible worlds. But surely *Rio de Janeiro*, as it is now, is not exactly as it was before: there are more people, more pollution and it is not anymore the capital of Brazil. But some people would say: Rio de Janeiro will always be Rio de Janeiro, as if there was something permanent, essential which characterizes the town, a kind of spirit of the city. But maybe it is the use of language itself, the use of a proper name that solidifies the variations. This proper name means nothing, then for this very reason its meaning does not vary, such kind of name is a fix point to something fuzzy.

The difference between proper names and definite descriptions is that in the case of proper names, it constitutes the basis of identity (two objects are considered as essentially the same). It is a *proper naming identification*. In case of definite descriptions it is just a feature in common (two recognizable different objects have a common feature), which can be essential or not. It is a *descriptive naming identification*. Definite descriptions are based on a more rational identification and proper names on a more creative one, not to say artificial one.

Let us now turn to naming differentiation. *The capital of Brazil* and *Brasilia* are two names for one thing in the present world, one name is a definite description, the other one is a proper name. *The capital of Brazil* is a name which refers to a city in function of a certain feature, this feature may be accidental. Why referring to a city in an accidental way? Because in some context this feature is the most important, it gives the adequate information, for example when we are saying: *The president of France is traveling to the capital of Brazil.*

Name differentiations may be based on ignorance, like the story of the morning star and the evening star. I can call a mouse Mini and another mouse Kiki and at the end of the day see that there are one and the same mouse. It can also be used to create a difference when there is not necessarily one: Norma Jeane Mortenson was also called Marylyn Monroe. The splitting of one person through the use of two proper names may lead to suicide or schizophrenia.

3 Naming identification and naming differentiation for plural terms

These two processes can also happen for plural terms. Note first that the distinction between proper names and definite descriptions can also be made for plural terms: *Inhabitants of Rio de Janeiro* is a definite description and *Cariocas* is a proper name to refer to the same people. So here we have an example of naming differentiation which works quite in the same way as for naming differentiation for singular terms.

To use a definite description in case of plural terms is a way to capture a collection of objects by one feature they have in common which can be more or less accidental, such as red objects, edible mushrooms or hot dogs. In this case we can speak about loose identity. But one may want to find a feature which captures an essential feature such as when we say that *human beings* are *rational animals*. The idea of rational animal is a way to unify a group of animals at the same time characterizing them, as if the identity was real. The identification process is then stronger than to use a proper concept (i.e. a concept working like a proper name) which is just a name without particular meaning, like *human beings*. Using a proper concept may be quite ambiguous because it is a way to artificially build an identity made of fuzzy features through a tag, like in nationalism: I am proud of being a Poldavian.

Let us now consider a limit case, the concept of *object* in the wide sense of the word, something which exists or can be thought: a cat is an object, anger is an object, Donald Duck is an object. They all are objects, what they have in common is to be an object, all are the same because all are objects. This is the opposite of individuality through the principle of identity according to which each object is identical to itself but different from the other ones.

4 Identification and differentiation in mathematics

It is important to stress that although language may have an important role in the process of identity, identification and differentiation may be considered beyond language and names. Identity can be seen as a relation between objects and/or between thoughts. This is not necessarily obvious for people who put language in the first place and have the idea that it is predominant in the thought process. This is generally the case of philosophers of language, who are studying thought through some analyses of language, like the distinction between definite descriptions and proper names. And it is also the case of modern logicians who have studied reasoning through the construction of formal languages.

But mathematics can be considered as a good example where naming identification and naming differentiation can be understood beyond a pure naming process, where identification and differentiation are performed beyond language.

As an example of identification, we can examine the case of *zero*: zero as a natural number and zero as a rational number are different, they have different properties. The reason why the same name is used is that they have also some common features which can be expressed in particular by the fact that the structure of natural numbers is a substructure of the structure of rational numbers. The notion of structure, which is the key notion of modern mathematics, permits to explain how identity works.

Some people may have the idea that a number is an entity with some mysterious features hidden behind a symbol, a proper name, like 7. But according to modern mathematics an object is nothing else that its relations with other objects in a structure, which also can be understood by its relations with other structures. To identify an object is to identify a structure, or better, a class of structures.

The identity of zero is its properties in a given structure or/and in a class of structures. Several solutions are possible. As we were saying we may want to identify the zero of the natural numbers with the zero of the rational numbers, but their identity here is based on algebraic characteristics with regards to multiplication and addition. Now from the point of view of order, zero is quite different as a natural number and as a rational number. It makes no sense to say that there is one true zero. The zero of the natural number maybe is the firstborn but it has grown and has many aspects: the zero of rational numbers, the zero of real numbers or something like the empty set — then another name is used, maybe not because its properties are so much different but rather because it is conceived from a different point of view. This could appear as a naming differentiation.

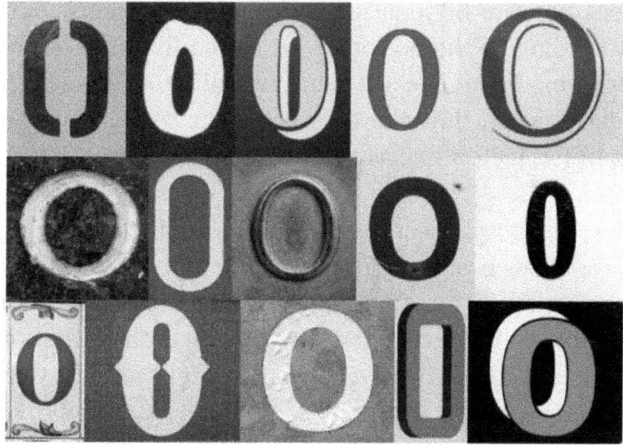

One may want to characterize precisely the zero of natural numbers. This can be done for example through Peano axioms of arithmetic, in first-order logic or in second order logic. As it is known, in first-order logic the theory is not categorical, there are several non-isomorphic models and zero is not the same in these different models. It is not the same for example in the standard model and in a non-standard model where there are other numbers after the standard natural numbers that are out of reach of zero through the successor function. In first-order logic, we are able to characterize zero with some axioms, however these axioms characterize not just one zero, but a multiplicity of zeros that are identified through the notions of axioms and models. Model theory is a way to understand the identification process. In second-order logic it is possible to have a categorical axiomatization of arithmetic, where all models are isomorphic (given a fixed cardinality). In this case zero is always the same up to isomorphism.

Isomorphism is identity between structures. It is the easiest way to identify structures. Now there is a more complex one which is equivalence between structures in the following sense: two structures are equivalent if they have a common expansion by definition up to isomorphism. An example of such equivalence is between an idempotent ring and a complemented distributive lattice. Equivalence between structures can be compared with translations between languages, which are not therefore not generally considered just as simple isomorphisms. If one language can be translated into another language and vice versa we can say that they are identical. In this sense the Bible in English can be considered as identical to the Bible in French.

Now let us see how (naming) differentiation works in mathematics. The differentiation process can be clearly explained using the concept of congruence: two objects are different but can be identified through congruence. For example $p\&p$ is congruent to p, according to the notion of logical equivalence which is a congruence in classical propositional logic. We are using two different names, "$p\&p$" and "p", to denote two different objects, that can be seen as the same. In the original structure they are different, in the factor structure, they are the same. Differentiation is based on identification.

In mathematics a structure where there are no non trivial congruences is called a simple structure. In such a structure two different objects cannot be identified according to a congruence. For example the structure of natural numbers is a simple structure: it is possible to identify 2 and 4 saying that they are both even numbers, but they cannot be identified through a congruence relation, they are really different.

5 Leibniz principle of identity and congruences

According to something which is generally called "Leibniz principle of identity", *two objects are identical if and only if they have the same properties*. This principle can be split in two: *identity of indistinguishables* (if two objects have the same properties they are the same) and *indistinguishability of identicals* (If two objects are the same, they have the same properties).

Indistinguishability of identicals is nothing else than congruence. In a given structure there is a whole hierarchy of congruence relations, the lattice of congruences: Leibniz identity is the strongest one, the principle of identity as we were presenting it at the beginning (*everything object is identical to itself but different from other objects*) is the weakest one. If we have a simple structure, Leibniz identity is the same as the principle of identity, this is the only congruence.

When a structure is not simple, *two different* objects can be the same in the sense of Leibniz. But one can say that the only difference is that they are two, not just one, we can speak about numerical difference in this case. An example is a structure with a two element set, Bibi and Titi, and with a universal binary relation: every object is related to every object. In this case Bibi and Titi are the same according to Leibniz identity. Leibniz identity is not so interesting, it is a trivial case like the identity given by the principle of identity. What is interesting is to identify things that are different in a sense which is not purely numerical.

This is what happens with non trivial congruences, which is probably the most precise formulation of what a subtle identity can be beyond Charybdis and Scylla. This notion is a central tool in modern mathematics and is directly connected with another key notion of modern mathematics: the notion of morphism - since a morphism generates a congruence and vice versa. The notion of congruence in its generality was presented by the General Bourbaki. Maybe only a Cretan warrior could find the way through Charybdis and Scylla.[1]

---

[1] Bourbaki's definition is not that simple and it is not necessarily easy to give a general definition of congruence because it depends on a general theory of structures, to develop such a theory is also a difficult task. This paper is basically non-technical, for a preciser treatment the reader may have a look at my 1996 and 1997 papers. I intend to develop the meaning of congruence relations outside of the realm of mathematics in a further paper. I would like to thank Daniel Parrochia for useful remarks and comments.

6 Dedication and personal recollections

I had many interests in commons with Dale Jacquette: Boole, Schopenhauer, the Square of Opposition, ... And like me Dale was working on many different topics. He moved to Switzerland shortly after I left the country. I visited him in Bern when I was working in Brazil. I gave a talk there at the Institute of Philosophy and we had dinner together, with his wife Tina and my wife Catherine. I remember that he told me that one of the reasons he left USA was because its department (at Penn Sate University) was dominated by continental philosophy. He took part to three editions of the series of congress I am organizing on the square of opposition: the 2nd edition in Corsica in 2010, the 3rd edition in Beirut in 2012 and the 4th edition in the Vatican in 2014. We published together a book resulting from the Corsica edition (Beziau-Jacquette 2012).

Dale Jacquette at the 3rd World Congress on the Square of Opposition American University of Beirut, Lebanon, June 2012.

He took also part to two editions of another series of events I am organizing, UNILOG: World Congress and School on Universal Logic. the 3rd UNILOG that took place in Lisbon in April 2010 and the 5th UNILOG organized in Istanbul in June 2015 where he gave a tutorial on Boole. The present paper deals with a topic Dale liked a lot: identity. His B.A honor thesis (1975) was on *Aristotle on identity and identification*. Along his carrier, this was a recurrent topic, see e.g. his 2011 paper "Frege on Identity as a Relation of Names". We posthumously published in the *South American Journal of Logic* his piece "Anatomy of a Nonidentity Paradox" (2016). I have myself been working on this topic since more than 20 years. This paper is the 10th in a series of paper I wrote on identity and I am glad to dedicate it to Dale.

References

[1] J.-Y.Beziau, 1996, "Identity, logic and structure", Bulletin of the Section of Logic, 25, pp.89-94.

[2] J.-Y.Beziau, 1997, "Logic may be simple", Logic and Logical Philosophy, 5, pp.129-147.

[3] J.-Y.Beziau, 1998, "Do sentences have identity ?", in Proceedings of the XXth World Congress of Philosophy, The Paideia Project, Boston.

[4] J.-Y.Beziau, 1999, "Was Frege wrong when identifying reference with truth-value?", Sorites, 11, pp.15-23.

[5] J.-Y.Beziau, 2003, "Quine on identity", Principia, 7, pp.1-15.

[6] J.-Y.Beziau, 2004, "What is the principle of identity? (Identity, Congruence and Logic)?", in Lógica - teoria, aplicaDŹes e reflexŻes, Feitosa and Sautter (eds), CLE, Campinas.

[7] J.-Y.Beziau, "Sentence, proposition and identity", Synthese, 154 (2007), pp.371-382.

[8] J.-Y.Beziau, 2009, "Mystérieuse identité", Le même et l'autre, identité et différence - Actes du XXXIe CongrŔs International de lÕASPLF, Eotvos, Budapest, pp.159-162.

[9] J.-Y.Beziau, 2015, "Panorama de l'identité", Al Mukhatabat, 14, pp.205-219.

[10] J.-Y.Beziau and D.Jacquette, Around and beyond the square of opposition, Birkhäuser, Basel, 2012.

[11] N.Bourbaki, 1939, Théorie des ensembles — Fascicule de résultats, Hermann, Paris,

[12] G.Frege, 1883, Über den Zweck der BegriffsschriftÕ, Jenaische Zeitschrift f§r Naturwissenschaft, 16 , pp. 1—10.

[13] W.Hodges, 1983, "Elementary predicate logic", in Handbook of philosophical logic I, D.Gabbay and F.Guenthner (eds), Reidel, Dordrecht, pp.1-131.

[14] D. Jacquette, 2011, "Frege on Identity as a Relation of Names", Metaphysica 12 (1):51-72.

[15] D. Jacquette, 2016, "Anatomy of a Nonidentity Paradox", South American Journal of Logic, Vol. 2, n. 1, pp. 119—125.

[16] D. Jacquette, 2016, "Subalternation and existence presuppositions in an unconventionally formalized canonical square of opposition", Logica Universalis, Volume 10, Issue 2-3, June 2016, pp.191-213.

[17] D.Krause and J.-Y.Beziau, 1997, "Relativizations of the principle of identity", Logic Journal of the Interest Group in Pure and Applied Logics, 5, pp.327-338.

[18] S.Kripke, 1980, Naming and necessity, Harvard University Press, Cambridge.

[19] W.V.O.Quine, 1960, Word and object, MIT, Cambridge, Mass.

[20] L.Wittgenstein, 1921, Tractatus Logico-Philosophicus.

Inside Außersein

FILIPPO CASATI
Kyoto University, Japan.
`filippo.g.e.casati@gmail.com`

GRAHAM PRIEST
CUNY Graduate Center, USA and the University of Melbourne, Australia.
`g.priest@unimelb.edu.au`

1 Introduction

In 1973, Gilbert Ryle famously declared that Meinong's theory was dead and buried (Ryle 1973). Now, after more than 40 years, history has shown that he was wrong. Meinong's *Theory of Objects* is more alive than ever: the proof is provided by the vast contemporary literature that engages seriously with many of the important ideas of Meinong. Nevertheless, a fundamental notion of Meinong's *Theory of Objects* has not yet received the attention that it deserves. The notion in question is that of *Außersein*. In this paper, we try to begin to fill this gap in the literature.

After introducing the main ideas of Meinong's philosophy (Section 2), we critically evaluate the few available interpretations of Meinong's *Außersein*. In particular, after showing that some contemporary Meinongians try to evade *Außersein*, we criticize the interpretations delivered by Grossman and Lambert (Section 3). Finally, in Sections 4 and 5, we defend the interpretations proposed by Jacquette and Priest. We compare them, drawing attention to both their similarities and differences.

2 Enter Meinong

Think about the last time you went to Paris. Remember your visit to Notre-Dame, the paintings you saw at the Centre Pompidou, the cheese you ate in that beautiful brasserie, and the picture you took in front of the Eiffel Tower. Maybe you have never been there. In that case, just imagine these things. When you remember or imagine something, you actually remember or imagine a *thing*. Paris, the paintings, the brasserie, and the Eiffel Tower are the things you remember or imagine.

If you share this intuition, you may want to agree with Meinong in assuming that usually intentional activities are directed towards a thing, and that, as such, they require something towards which they are directed. In Meinong's words: "psychological events... have this distinctive character of being directive towards something" (Meinong 1904, pp. 76 – 77).[1] This is why "knowing is impossible without something being known, ...judgements and ideas or presentations are impossible without being judgements about and presentation of something" (Meinong 1904, p. 76). In Meinong's terminology, the things towards which our intentional activities are directed are the objects of our thought—*Gegenstände*.

Of course, remembering and imagining are only two amongst the many possible intentional activities we can engage in. Examples of different kinds of intentional activities are: worshiping, fearing, praying to, and panicking about. Moreover, given so many different kinds of intentional activities, it is natural to expect that we can direct them towards very many different kinds of objects as well. In particular, we can worship or fear objects that exist, and we can worship or fear objects that do not exist.

This variety is not lost in Meinong's theory of objects. Indeed, Meinong believed that some objects have being [*Sein*] and some objects do not. Objects that have being can either exist or subsist [*bestehen*]. When they exist, such as the Eiffel Tower, they are either spatially or temporally located (or both); when they merely subsist, such as prime numbers, they somehow exist without being spatio-temporal located.[2] Objects that do not have any kind of being, such as purely fictional characters, have non-being [*Nichtsein*]. In the Meinongian framework, being (both existence and subsistence) and non-being are simply normal properties. Of course, the Eiffel Tower, prime numbers, and fictional characters are very different kinds of objects; still, according to Meinong, they are objects. Since we can grasp them with our intentional activities, they are all *Gegenstände*.

It is also important to recall that, according to a common interpretation (see, for instance, Berto 2013, pp. 85 – 109), Meinong thinks that every object, regardless of its ontological status, has the properties that it is described as having. For instance, even if Sherlock Holmes does not have any kind of being (indeed, he has non-being), he has the property of *being a detective* because, according to Doyle's stories, he is described as such. This idea is captured by the so-called naïve characterization

---

[1] Page references to quotations from Meinong are from their English translations.

[2] The relationship between existence and subsistence in Meinong's philosophy has been understood in different ways. Following what Meinong claims in his *On Assumptions* (1910, p. 59) and in his *Selbstdarstellung* (1921, p. 18), we assume that, according to Meinong himself, when an object exists, it subsists too. Even though this is certainly a simplification, this is also enough for our purposes. Indeed, no arguments defended in our paper rely on this matter.

principle, according to which any object has the properties that it is characterised as having. In other words, for any set of properties (for any *Sosein*), an object which satisfies that set of properties (that *Sosein*) is in the domain of discourse.

Let us call this view the *naïve* Characterisation Principle (CP). It is not tenable, for a number of reasons. The first is that the naïve CP delivers violations of the principle of non-contradiction admitting inconsistent objects, namely objects with inconsistent properties. The second is that the naïve CP allows us to prove the existence of whatever we want. Third, and most generally, the naïve CP delivers triviality, in the technical sense that any sentence follows.[3]

In order to avoid these predicaments, both Meinong and neo-meinongians have rejected the naïve CP. There are a number of ways of doing this. Two will be particularly important in what follows, so let us spell these out here. The first, *nuclear* Meinongianism, divides properties into the nuclear (or characterising) and the non-nuclear (non-characterising). The CP applies only when the properties involved are nuclear. This version is endorsed, for example, by Parsons (1980). On another version, *modal* Meinongianism, any properties can be deployed in the CP; but the object characterised is not guaranteed to have the characterising properties at the actual world (though it may). It is guaranteed to have them at *some* world—maybe an impossible world. This version is endorsed by Priest (2005) and Berto (2013). We do not need to go into the question of which, if either, of these versions is correct. It suffices here to say that both versions agree that objects, whether they exist or not, can have properties in a perfectly ordinary sense.

In summary, then, we can say that Meinongians are committed to the following philosophical picture. Anything one can refer to is an object (*Gegenstand*), X. Since X is an object, it has properties. Since X has properties, X has a *Sosein* comprising these properties. So much, at least is common ground.

We now turn to where the common ground disappears; and it does so with the notion of *Außersein*. Meinong held that an object may have *Außersein*; but what, exactly, does that mean? That is the topic of the rest of this essay. In the next section, we will review and critically assess some of the most important interpretations of *Außersein*. In the following section, we will look at the interpretation delivered by Jacquette and Priest, which we take to be the most coherent interpretation with respect to Meinong's ontology. In the following section, we will explore the most important difference between Jacquette and Priest on the matter of *Außersein*.[4]

---

[3]See Priest 2005, 4.2.

[4]The authors refer to Priest in the third person, not because his present self wishes to distance himself from his prior self, but just to avoid clunky syntax.

3 Meinong's *Außersein*

3.1 Evading the Issue

A common attitude towards *Außersein* is one of avoiding it.[5] People endorsing this approach sometimes focus on the so-called *Principle of Außersein* without discussing what *Außersein* actually is. Meinong held that the *Sosein* of an object was independent of its *Sein*. This principle, the "principle of the indifference of pure Objects to being" (Meinong 1904, p. 86), is sometimes called the *Principle of Außersein*. Understood in this way, the Principle, then, enunciates one of the core idea of Meinong's theory; and of course, understood in this way it is extremely important. Nevertheless, it seems a mistake to suppose that this says anything about *Außersein* as such. Let's see why.

First, very many different disciplines employ principles which are supposed to be theoretically helpful: for instance, statistics makes use of principles to better understand probability, physics to better understand gravity, and philosophy to better develop metaphysical concepts. However, generally speaking, though the 'principle of X' concerns X, it is not itself X. This is true for the Principle of *Außersein* as well. As the Newtonian law of gravity is concerned with gravity without being gravity, the Principle of *Außersein* is concerned with *Außersein* without being *Außersein*. Even though the Principle of *Außersein* can be intuitively related to *Außersein*, these two concepts are still different. Meinong himself distinguishes between the principle which is concerned with the *Außersein* of a 'pure object' [*Der Satz vom Außersein des reinen Gegenstandes*], on the one hand, and *Außersein* itself, on the other hand.[6].

Moreover, one should not forget that, according to Meinong, every time there is an intentional act, the act is at last normally directed towards an object. If so, *Außersein* has to be an object as well. Indeed, not only are we talking and thinking about *Außersein* right now, but Meinong himself does so too. From this point of view, since it is legitimate to investigate what any object of thought is (be it Sherlock Holmes, the Eiffel Tower, or a prime number), it is legitimate to investigate what *Außersein* is as well. Therefore, this evasive account does not get to the heart of the matter.

---

[5]Examples of philosophers displaying this attitude towards *Außersein* include Berto (2013) and Parsons (1980).

[6]For instance, in his *The Theory of Objects*, Meinong talks about the Principle of *Außersein* (1904, p. 86) and, in his *On Emotional Presentation*, he explicitly refers to *Außersein* itself (1917, p. 15).

3.2 Grossman's Position

An alternative approach to the notion of *Außersein* is proposed by Grossman (2008). For him *Außersein* is what is outside being and non-being.[7] In order to see what this means, consider an object A, which has being. According to Grossman, there is no such thing as object A with its own being because, following Meinong, A is nothing more than A itself. There is no addition of A's being to A itself. Further, consider an object A with non-being. Once again, there is no such thing as A with its own non-being because A is nothing more than A itself. There is no addition of A's non-being to A itself. Hence, to say that A has *Sein* (existence, subsistence) is not to say that being is part of A. In the same way, to say that A has *Nichtsein* is not to say that non-being is part of A. According to Grossman, "[existence and subsistence] cannot be part of objects" because, since they are literally outside being, "there are no such entities [or objects] as existence and subsistence" (Grossman 2008, p.119). For the same reason, non-being cannot be part of an object because there is no such object as non-being. Grossman is also clear that 'being a part of an object' means 'being an ontological constituent of an object'. He writes that something is part of an object "in the way in which properties (and relations) or instances of properties (and relations) are parts of objects" (Grossman 2008, p.119). In other word, *being red* is part of a red object because it constitutes the redness of the object in question.

To see why Grossman's interpretation is implausible, let's start by noting that the expression "there are no such entities [or objects] as existence and subsistence" is ambiguous. To begin with, it can be interpreted as 'existence and subsistence are not entities [or objects]'. If this is the case, then Grossman's interpretation of *Außersein* is incompatible with Meinong's account of intentionality, according to which every time we refer to something, we refer to an object. Since we can refer to existence and subsistence (since, for instance, we can say that 'existence is a property' and that 'subsistence is different from existence'), they have to be objects too. The same holds for non-being. Since Meinong's account of intentionality is crucial for his whole philosophical project, it is not plausible that Meinong abandons it when he deals with being and non-being. For this reason, this interpretation of Grossman does not look promising at all.

A second possible interpretation of 'there are no such entities [or objects] as existence and subsistence' interprets it as 'entities [or objects] like existence and subsistence do not have being' — and the same for non-being. In Meinongian terms, this means that neither being nor non-being has being: they neither exist nor subsist.

---

[7] As Routley has pointed out in some unpublished notes (Sylvan, 1950), Grossman's interpretation of *Außersein* is unclear. We agree with Routley. What follows is the best we can do to make sense of it.

In this sense, being and non-being are *Außersein* — they are literally outside being. As strange as it may look, this is a position that Meinong can, at least theoretically, endorse. At the end of the day, it could be the case that, according to Meinong, being and non-being have the same ontological status as fictional characters.

However, this interpretation is still highly problematic. For in this case, according to Grossman, being and non-being cannot be part of an object exactly because they neither exist nor subsist. Now, as we have already noted, according to Meinong, being and non-being are properties. If being and non-being neither exist nor subsist, it is natural to think that no other properties exist or subsist either. There is no reason to believe that being and non-being constitute exceptions. Moreover, if being and non-being cannot be part of an object exactly because they neither exist nor subsist, then no other properties can be part of an object because they neither exist nor subsist either. This means that, for instance, the properties *being red* and *being sweet* can not be part of a red sweet object; and that, given Grossman's definition of 'being part of', they cannot constitute the redness or the sweetness of the object in question (Grossman 2008, p.119).

If what we have said until now is correct, this second understanding of Grossman's interpretation of *Außersein* contradicts the principle of *Außersein*, namely Meinong's fundamental assumption that, regardless their ontological status, objects have the properties that they are characterised as having. Hence, Grossman's account of *Außersein* contradicts Meinong's view that objects always have a *Sosein* composed of the properties that they are characterised as having. Even in this case, Grossman's understanding of *Außersein* does not seem correct.

3.3 Lambert's Position

A third account of *Außersein* is proposed by Lambert in his *Meinong and the Principle of Independence* (1983). According to Lambert, *Außersein* is the domain of all objects without being. "The domain of nonbeings Meinong called *Außersein*", which is, literally, "the domain of objects outside of being" (Lambert 1983, p. 14). As he points out, such a domain is enormous and, among its denizens, it contains "possible objects such as Pegasus or the golden mountain and also impossible objects such as the round square of Mill and the proof of the decidability of general quantification theory" (Lambert 1983, pp. 14 – 15). In other words, *Außersein* is understood as the set of all objects that do not have either the property of *being existent* or the property of *being subsistent*.

Unfortunately, as for the other interpretations discussed so far, this one faces problems as well. For a start, according to the Lambert, *Außersein* is a domain. If so, as an element of a domain does not *have* or *possess* a domain but it is *member of* a domain, objects should not have or possess *Außersein* but they should be a member of *Außersein*. Unfortunately, this is not what we read in Meinong. Meinong himself does not talk about domains; neither does he use any terminology that seems to support the identity between *Außersein* and domain.

More importantly, according to Meinong, *all* objects have *Außersein*, not just objects with *Nichtsein*.[8] Thus, in his *On Emotional Presentation*, he writes: "If an ... object is to be apprehended, this object ..., at least as having *Außersein*, must be given as a precondition for the experience" (Meinong 1917, p. 15); and *"Außersein* seems clearly to be predicable of all objects" (Meinong 1917, p. 19). So Lambert's interpretation turns Meinong into a nihilist: every object has *Nichtsein*. That is, no object exists. This conclusion is evidently against Meinong's view, according to which, even though some objects do not have being, some other objects do.

4 Jacquette and Priest's Position

So let's move on to the last account of *Außersein* we will consider, namely the view that assimilates the notion of *Außersein* to the notion of objecthood. Such an interpretation we take to be the most coherent among the ones available on the market, and it has been recently developed in two different ways by Dale Jaquette, in his *Alexius Meaning: the Shepherd of Non-Being* (2015), and by Graham Priest in his '*Sein* Language' (2014b).

Let's start with Jacquette. As does Lambert, he takes *Außersein* to be a domain of objects. However, contrary to Lambert's interpretation, Jacquette believes that *Außersein* is not a notion concerned with any kind of ontological issues: it is not about objects without being, and is not about objects with being either. For this reason, *"Außersein* is not a special kind of *Sein* [and it is not a special kind of *Nichtsein* either]": "[it] is not a subcategory of the ontology" (Jacquette 2015, p. 71). Therefore, Jacquette interprets *Außersein* as the domain of all objects, regardless of whether they exist, subsist, or neither exist nor subsist. *Außersein* is simply "the name Meinong later gives to what ... [is] considered independently of its ontic status" (Jacquette 2015, p. 71). According to Jacquette, all objects belong to *Außersein*; as such, *Außersein* is intended as "an ontologically neutral referential domain" (Jacquette 2015, p. 71). In this sense, the notion of *Außersein* is tightly connected with the notion of objecthood: all objects are members of *Außersein* exactly because all objects are (trivially) objects. They all belong to *Außersein* in virtue of their objecthood.

At this point, one may be suspicious of Jacquette's interpretation because, as we

---

[8]Cf. Meinong, 1904, p. 83-86; Meinong, 1917, p. 19; Marek, 2013.

have argued in the case of Lambert, it would be more natural to think of *Außersein* as a property that all objects have, rather than a domain in which all objects are. This worry disappears if we move from Jacquette's interpretation to Priest's interpretation (2014b). Indeed, according to this, Meinong's idea of *Außersein* is nothing more than the property of *being an object*. As such, every object has *Außersein* because (trivially) every object is an object. He writes: "Any object has *Außersein*. That is, it is simply an object" (Priest 2014b, p. 439). In other words, if something is an object, it has *Außersein*; and if something has *Außersein*, it is an object.

It is interesting to note briefly that, in his *Exploring Meinong's Jungle and Beyond* (1980), Richard Routley seems to agree with Priest's interpretation. He writes that "An object as such is said to *be ausserseiend* or to *have Aussersein*. That is, *Aussersein* is a property ... ; it is the property of objects as such, such that existence and non-existence are external to them" (Routley 1980, p. 857). As does Priest, Routley takes *Außersein* to be the property that all objects have as objects, as such; that is the property in virtue of which objects are simply objects, regardless of their ontological status. In other words, *Außersein* is taken to be the property of *being an object*. Unfortunately, this observation is isolated and, after that, Routley focuses his attention exclusively on the Principle of *Außersein*. So let's go back to Priest.

From the idea that *Außersein* is the property of *being an object*, Priest infers that *Außersein* is a metaphysically fundamental property—an *ur*-property; that is, a property the instantiation of which is entailed by any other property. In order for something to be red, green, tall, or heavy, this something has to be a thing, an object, in the first place. Having *Außersein* is therefore a property necessarily entailed by any other property.

Recall, also, that in Meinong's framework, everything has a *Sosein*, a certain collection of properties. Given this, when something has *Außersein* (when something is an object), it has properties too (it has a *Sosein*). Since *Außersein* is a property, namely the property of *being an object*, even an object which has no property other than *Außersein*, has a *Sosein*. This is why Priest claims that "*Außersein* [and] *Sosein* are equivalent" (Priest 2014b, p. 439).

Contrary to all the other understandings of *Außersein* we have discussed above, both Jacquette's and Priest's interpretations have the unquestionable advantage of being completely consistent with Meinong's expressed views. Since Meinong claims that all objects have *Außersein*, but only some objects have being and only some objects have non-being, it is natural to think that there is something, namely *Außersein*, which is, so to speak, shared by all objects regardless their ontological status. For Jacquette, this is being in the domain of all objects; while for Priest, it is the property of *being an object*.

Meinong himself seems to endorse this idea when he says that every object, "has a remnant of a positional character, [that is] *Außersein*" (in Grossman 2008, p. 228). Here, Meinong suggests that, regardless of the ontological status of an object, there is always something contributing to its 'positional character', namely something that makes the object an object 'possibly present' to the consciousness of a subject. Now, in the Brentenian tradition of which Meinong is a part, *being possibly present to a consciousness of a subject* simply means *being an object*. As such, *Außersein* is what makes an object an object. And this is exactly the role played by the notion of *Außersein* in both Jacquette and Priest.

So much for the similarities between Jacquette's and Priest's views. Now for the differences. One of these, we have already commented on: the fact that Jacquette takes *Außersein* to be a set, whereas Priest takes it to be a property.

Next, Priest ties in the notion of *Außersein* with that of identity, since something is an object if and only if it is self-identical (*Identitätsein*) (Priest 2014b, p. 439). As we have noted, for Priest, *Außersein* and *Sosein* are equivalent. Moreover, everything that is an object has the property of being self-identical, and vice versa. *Außersein*, *Identitätsein*, and having a *Sosein* all, then, come to the same thing.

By contrast, Jacquette does not mention identity at all. What his views are on the matter we don't know. But perhaps he would be quite happy to accept the connection between *Außersein* and identity. Arguably, the set of objects and the set of things that are self-identical are the same set.

Perhaps the biggest difference between Priest and Jacquette concerns their treatment of the Characterisation Principle. Jacquette is a nuclear Meinongian, whilst Priest is a modal Meinongian.

According to Jacquette, properties can be divided into *constitutive* and *extra-constitutive* properties; this is basically the same distinction as Parsons (1980)'s distinction between nuclear and extra-nuclear properties. Constitutive properties are properties which are taken to be essential to determining the nature of an object, while extra-constitutive properties are properties that are implied by the constitutive ones (Jacquette 2015, p. *xxx*). Jacquette then holds a version of the CP according to which the set of properties which characterise an object, namely the object's *Sosein*, can contain only constitutive properties. He writes: "*Außersein* of the pure object is the referential semantic domain of all objects understood only as objects, constituted in their *Soseine* exclusively by their distinguishing constitutive property clusters, without taking their ontic status into account" (Jacquette 2015, p. 71).

Priest, on the other hand does not endorse a version of CP which relies on the distinction between constitutive and extra-constitutive properties. Endorsing modal Meinongianism, Priest believes that an object can have *all* the properties it is characterised as having, either at the actual world, or at some other (possible or impossible) world. Thus, any way of characterising an object, will determine an object, that is, something with *Außersein*. The characterisation, however, is not guaranteed to be true of the object at the actual world (though it may be) — just at some worlds.

5 Objects that are not Objects

This last difference has important ramifications for the notion of *Außersein*. In this section, we will look at these.

In the first place, the property of *being a member of Außersein* is not a characterizing property for Jacquette. Here, characterizing properties are properties that are essential to distinguish between objects (see Jacquette 2015, Ch. 5). Since all objects are members of *Außersein*, the property of *being a member of Außersein* does not help to distinguish between objects. As such, it cannot be a characterizing property either. And certainly, not-having-*Außersein* is not a characterising property. If it were, we could characterise an object, x, by the condition of not having *Außersein*. It would then not have *Außersein*; but since it is an object, it would have *Außersein* as well. However, Jacquette is no dialetheist.[9] So this would be quite unacceptable to him. For him, the domain of *Außersein* is quite consistent.[10]

Priest, on the other hand, is a dialetheist; but, as he is often at pains to point out, there is nothing in modal Meinongianism as such, that requires this. In particular, one can characterise an object, x, by the condition of not having *Außersein*. This is guaranteed to be an object, and so have *Außersein*; but it is not guaranteed not to have *Außersein* at the actual world — only at some world, w, or other. This is no more contradictory than Priest being a man at this world, and a woman (not a man) at some other. There is not even a reason to believe that x *is* an object at world w. It may well be a logical truth that everything is an object, that is, has *Außersein*. However, logical truths may fail at impossible worlds. So it may not even be true at w that x is an object.

One could, of course, characterise an object, y, as both having and not having *Außersein*. This does not mean that y is actually a contradictory object. All it means is that there is some world, w (and it would be natural to think that w is an

---

[9] It is difficult to find a quotation in which Jacquette explicitly rejects dialetheism. However, in none of his work is there a trace of accepting any contradiction as true. For this reason, it is fair to assume that Jacquette was not sympathetic with dialetheism.

[10] As a referee correctly noted, related issues are discussed by Meinong himself in relation with the so-called 'Defective Objects' (Meinong 1917). However, due to the complexity of Meinong's account of defective objects, we do not discuss the matter here. A detailed discussion of defective object can be found in Casati and Fujikawa (draft).

impossible world) where y has these contradictory properties.

Having said all this, because Priest is a dialetheist, it is open to him, in a way that it is not open to Jacquette, to hold that some objects both do and do not have *Außersein*.

In fact, Priest does hold this view. He holds that nothingness both is and is not an object. It is clearly an object, since one can refer to it, think about it, etc. But it is not an object. By definition, it is the *absence* of all things: it is what remains when all objects are *removed*.

In fact, the contradictory nature of nothingness, even if were not obvious, can be proved with the help some simple machinery machinery.[11] First, according to Priest, to be an object is to be something. So let us define x *is an object*, Gx ('G' for *Gegenstand*), as $\mathfrak{S}y\, y = x$. As is clear, it is a logical truth that everything is an object: $\mathfrak{A}xGx$. That is, nothing is not an object: $\neg\mathfrak{S}x\neg Gx$.[12]

Next, we need a little mereology. Let us write $x < y$ to mean that x is a proper part of y. (Nothingness and y are the two improper parts of y.) As usual, $x \leq y$ means that $x < y \vee x = y$. x overlaps y, $x \bigcirc y$, can be defined in the usual way:

[1] $x \bigcirc y \leftrightarrow \mathfrak{S}z(z \leq x \wedge z \leq y)$

The sum, or fusion, of a bunch of objects is the object one obtains by putting all the objects together. Thus, the sum of your parts is you. So, given a bunch of objects, something will overlap their fusion iff it overlaps one of them. Let us write the sum of the things that satisfy the condition $A(x)$ as $\sigma x A(x)$. Then we have:

[2] $x \bigcirc \sigma x A(x) \leftrightarrow \mathfrak{S}y(A(y) \wedge x \bigcirc y)$

Now, whether or not every bunch of objects has a sum is a philosophically contentious matter. Some hold that they do; some hold that there is no sum if the objects do not hang together in an appropriate fashion (like the parts of a body or a country). We need take no stand on this matter here.

We are now in a position to define nothingness, n. It is simply the sum of no things, that is, no objects. (Thus, we might say that it is the sum of all the things in the empty set). Hence:

[3] $n = \sigma x \neg Gx$

This is the intuitively correct definition of nothingness. Moreover, the things that are not objects cannot fail to hang together (whatever that means) since there are none of them. Hence, [2] and [3] give us:

---

[11]For what follows, see Priest 2014a.

[12]Following Priest (2005), we write $\mathfrak{S}$ and $\mathfrak{A}$ as the particular and universal quantifiers, respectively, to bring home the fact that they are not "existentially loaded".

[4] $x \bigcirc n \leftrightarrow \mathfrak{S}y(\neg Gx \wedge x \bigcirc y)$

Clearly, Gn. (That, as we observed, is a logical truth.) But we can now show that n is not an object by the following simple argument. We know that $\neg \mathfrak{S}x \neg Gx$, so [4] gives us:

- $\forall x \neg x \bigcirc n$

and so:

- $\neg n \bigcirc n$

but then [1] gives us:

- $\neg \mathfrak{S}z\, z \leq n$

In particular, then:

- $\neg n \leq n$

and so:

[5] $n \neq n$

Now, for any x, either $x = n$ or $x \neq n$. In the first case, [5] and the substitutivity of identicals gives us $x \neq n$. So $x \neq n$ in either case. That is:

- $\mathfrak{A}x\, x \neq n$

i.e.:

- $\neg Gn$

Hence we have $Gn \wedge \neg Gn$: n is an object that is not an object.

What we see, then, is that for Priest *Außersein* is an inconsistent notion: even though everything has it, some things do not (as well).

6 Conclusion

Let us conclude by summarizing the main points of this paper. First of all, distinguishing between the Principle of *Außersein* and *Außersein* itself, we have shown that, even though contemporary Meinongians are engaged with the former, they often ignore the latter. Secondly, we have examined the few interpretations of *Außersein* available on the market. On the one hand, we have shown that both the interpretations defended by Grossman and Lambert are incompatible with the general framework presented by Meinong in his *Theory of Objects*; on the other hand, we have defended both Jacquette and Priest's account of *Außersein*, according to which Meinong's *Außersein* is deeply related with the notion of objecthood. Finally, we have focused our attention on three main differences between these two interpretations, showing that: (i) for Jacquette, *Außersein* is a domain while, for Priest, it is a property; (ii) for Jacquette, there is no explicit connection between *Außersein* and self-identity while, for Priest, there is; (iii) for Jacquette, *Außersein* is a consistent notion while, for Priest, it is not: some objects do and do not have it. Adjudicating this last disagreement would, of course, take us a long way beyond the ambit of this paper.

References

[1] Berto, F. (2013), *Existence as a Real Property: The Ontology of Meinongianism*, London: Synthese Library, Springer.

[2] Casati, F. and N. Fujikawa, (Draft) On Defective Objects.

[3] Grossman, R. (2008), *Meinong. The Arguments of Philosophers*, London and New York: Routledge.

[4] Jacquette, D. (2015), *Alexius Meinong: the Shepherd of Non-Being*, London: Synthese Library, Springer.

[5] Lambert, K. (1983), *Meinong and the Principle of Independence. Its Place in Meinong's Theory of Objects and its Significance in Contemporary Philosophical Logic*, Cambridge: Cambridge University Press.

[6] Meinong, A. (1904), 'Über Gegenstandstheorie', translated in Chisholm, R. M. (eds.), *Realism and the Background of Phenomenology*, New York and London: The Free Press Collier-Macmillan, 1960.

[7] Meinong, A. (1910), 'Über Annahmen', translated as *On Assumptions*, Berkeley, Los Angeles and London: University of California Press, 1983.

[8] Meinong, A. (1917), 'Über Emotionale Präsentation', translated in Schubert Kalsi, M. L., (ed.), *On Emotional Presentation*, Evanston: Northwestern University Press, 1972.

[9] Meinong, A. (1921), 'Selbstdarstellung' translated in Grossmann R., *Meinong*, London: Routledge Kegen Paul (1974).

[10] Parsons, T. (1980), *Nonexistent Objects*, Yale: Yale University Press.

[11] Priest, G. (2005), *Towards Non-Being*, Oxford: Oxford University Press; 2nd edn, 2016.

[12] Priest, G. (2014a), 'Much Ado about Nothing', *Australasian Journal of Logic* 11: 146-58.

[13] Priest, G. (2014b), '*Sein* Language', *The Monist* 97: 430-42.

[14] Routley, R. (1980), *Exploring Meinong's Jungle and Beyond*, Canberra: Australian National University RSSS.

[15] Ryle, G. (1973), 'Intentionality and the Nature of Thinking', *Revue Internationale de Philosophie* 27: 255-65.

[16] Sylvan, R. (1950), Papers 1950 – 1996, Box 23, Queensland: University of Queensland Library.

Truth and Interpretation

Dagfinn Føllesdal
Stanford University, USA and Universitetet i Oslo, Norway.
`d.k.follesdal@ifikk.uio.no`

Dale Jacquette contributed abundantly to a remarkably wide range of philosophical themes. He wrote or edited more than thirty books ranging from his dissertation on Meinong with Roderick Chisholm, over Schopenhauer, Wittgenstein, Hume, various ethical issues, including journalistic ethics, to *Frege: a Philosophical Biography*, forthcoming at the Cambridge University Press. And his more than 300 articles ranged even wider. He had much more to contribute. He was in the midst of very active research project on the ontology of music when he suddenly and unexpectedly died on 22 August 2016.

His broad orientation together with his sense of quality made him a very valuable member of the board for the Lauener Foundation, where we worked together during the last years of his life. I appreciated especially his openness. He never used labels like "analytic" and "continental" philosophy, which tend to close peoples mind. For him, quality was what counted, and reading this wide range of philosophers with an open mind, he learned that they could often have benefitted from reading one another, and that by bringing them together we could get insights that they had missed.

He was particularly engaged by ontological issues and wrote a general book on the topic: *Ontology* (2002). In two other books, *Meinongian Logic* (1996) and *Alexius Meinong, The Shepherd of Non-Being* (2015) he worked out his own view, inspired by Meinong's but freed from many of the problems and unclarities that afflict Meinong's conception Traditionally, the problems he deals with by appeal to objects are treated by help of a notion of meaning. This notion is notoriously unclear. However, this unclarity is matched by similar problems of individuation of objects in Meinong's theory. And meaning is needed for many other purposes, for example in order to account for communication and interpretation.

In what follows, I will focus on meaning, and emulate Dale Jacquette by disregarding the labels "analytic" and "continental." I have selected Gadamer, Davidson and Quine, three philosophers who focused on meaning, and where we can gain insight by bringing them together.[1]

Hans-Georg Gadamer's *Truth and Method*[2] is usually regarded as the breakthrough of the so-called "New Hermeneutics."

I shall first mention a couple of earlier main steps in the development of hermeneutics. After some remarks on the hermeneutic circle and the relation between hermeneutics and the natural sciences I will go on to discuss Gadamer's new approach to hermeneutics and its roots in Husserl's phenomenology. I will then present briefly two main approaches to meaning in so-called "analytic" philosophy, those of Quine and Davidson, and indicate their close connection with Gadamer's view on truth and method.

Reflections on interpretation started with the Greeks, particularly concerning the interpretation of Homer. The problems of interpretation become particularly acute where one is dealing with texts the correct interpretation of which is a matter of some importance, and which were written at a time or in a situation that are very different from those of the interpreter. Legal and religious texts are notable instances. Thus law and theology were the areas where hermeneutics was first systematically studied. In these two fields, interpretation was studied and debated throughout antiquity, into the Middle Ages and modern times.

A first major expansion of the field of hermeneutics came two hundred years ago, when Friedrich Schleiermacher (1768–1834) established hermeneutics as a separate discipline and expanded the area of texts to be studied to include literary and philosophical texts in addition to the legal and religious ones.

The next major expansion of the scope of hermeneutics came with Wilhelm Dilthey (1833–1911), who argued that hermeneutics applies to all "manifestations of the human spirit." Dilthey also endorsed the explanation-understanding thesis that had been put forth in 1858 by the historian Johann Gustav Droysen (1808–84) to the effect that while the natural sciences aim at explanation, the humanities and in part also the social sciences aim at understanding. The difference presumably

---

[1] Gadamer, by the way, was Wolfgang Künne's *Doktorvater*. Künne and his students are outstanding examples of how much one can benefit from studying philosophy with a broad perspective, not impeded by "schools" and labels. Two articles by Künne, where he anticipates part of what I am going to say here, are his inaugural lecture in Hamburg in 1981 on "Verstehen und Sinn. Eine sprachanalytische Betrachtung." *Allgemeine Zeitschrift für Philosophie* 6 (1981) 1, pp. 1–16, and "Prinzipien der wohlwollenden Interpretation", in: *Intentionalität und Verstehen* / hrsg. vom Forum für Philosophie Bad Homburg. — Frankfurt/Main: Suhrkamp, 1990, pp. 212Ð236. Parts of what I am going to say I have presented in "Meaning and experience," in Samuel Guttenplan, ed., *Mind and Language: Wolfson College Lectures 1974*. Oxford: Oxford University Press, 1975, pp. 34–35, and also in "Hermeneutics and the hypothetically-deductive method." *Dialectica* 33 (1979), pp. 319–336, and various other articles on hermeneutics.

[2] Hans-Georg Gadamer, *Wahrheit und Methode*. Tübingen: J.C.B. Mohr, 1960.

turned on whether what was being sought were causal laws or an elucidation of meaning. However, many different issues have been lumped together in this discussion and it is far from clear what the points are and how the arguments run. In any case, hermeneutics came to be conceived as the method of the humanities. Neither Droysen nor Dilthey said much about the method of the natural sciences.

In 1968 Jürgen Habermas (1929-) put forward the view that the natural sciences are characterized by the use of the hypothetico-deductive method, and thereby contrast with the humanities, which use the hermeneutic method, and the social sciences, which use what Habermas calls the "critical" method.[3] This is not the place to discuss this view. However, it would seem to count against it that hermeneutics shares the two defining feature of the hypothetico-deductive method: (1) setting forth interpretational hypotheses and (2) checking whether they together with our beliefs imply consequences that clash with our material, that is, with the text we are interpreting. Examples of this abound in the interpretation of literary works. Wolfgang Stegmüller showed this in 1979 in his interpretation of Walter von Der Vogelweide.[4] Another example is some passages in Ibsen's play *Peer Gynt* for which half a dozen interpretations have been proposed, most of which have afterwards been rejected because they fit poorly in with the text.[5] Rather than contrasting hermeneutics and the hypothetico-deductive method, I regard hermeneutics as the hypothetico-deductive method applied to meaningful material in order to bring out its meaning. I believe that Habermas has given up this view, but I have not seen any place where this is confirmed in writing.

The hermeneneutic circle

In hermeneutics, as in the natural sciences, we go back and forth between the hypotheses and the material until we achieve a fit, or "reflective equilibrium", to speak with Rawls.[6] We may find hypotheses that fit in with part of the material, but which have to be revised because they do not fit in with other parts. A good hypothesis must fit the whole material, and so will have to be modified until we find

---

[3] Jürgen Habermas, Erkenntnis und Interesse. Frankfurt: Suhrkamp, 1968.

[4] Wolfgang Stegmüller, "Walther von der Vogelweides Lied von der Traumliebe und Quasar 3 C 273". In Stegmüller, *Rationale Rekonstruktion von Wissenschaft und ihrem Wandel*. Stuttgart: Reklam Verlag, 1979 (Reklam Universal-Bibliothek 9938), pp. 27–86.

[5] See my "Hermeneutics and the hypothetico-deductive method." *Dialectica* 33 (1979), 319–336.

[6] John Rawls, *A Theory of Justice*. Cambridge, Mass.: Harvard University Press, 1971, p. 20 and several other places. The idea is old and was given especially clear expression in Nelson Goodman's *Fact, Fiction and Forecast*. Cambridge, Mass: Harvard University Press, 1955, pp. 65–68.

an interpretation that fits all the parts.

During this process the material itself against which the hypotheses are tested changes; passages that originally were interpreted one way come to be interpreted in another. This malleability of the material is more pronounced in hermeneutics than in natural science, but it has long been known there, too, under the title "the theory-ladenness of observation." Going back and forth between part and whole in this way is one ingredient in the renowned "hermeneutic circle." This expression was introduced in 1808, by the German theologian and philosopher Friedrich Ast (1778–1841). It had been noticed far earlier that when we interpret a text, our initial interpretation of a passage may come to change when we read it within the wider context of the whole text. And conversely, our interpretation of the whole text depends upon our interpretation of the parts. The hermeneutic *whole-part circle* includes, however, more than the text itself, for the text has to be understood within a context that comprises other works by the author, and also both its linguistic and its cultural setting. The setting helps us understand the text; the text, on the other hand, may help us see the setting in a new light, which in turn may change our interpretation of the text, etc.

Further, there is a *question-answer circle*: when we approach a text, we approach it with certain questions that may come to change as we get a better understanding of the text. These new questions, in turn, may change our interpretation of the text.

Finally, there is the *subject-object circle*: the totality that comes into play when we interpret a text comprises not only the text and its linguistic and cultural setting, but also us, the interpreters. We come to the text not only with explicit questions, but also with our whole horizon of beliefs and attitudes. Most of them we do not know, and are not even aware of, and many of them become changed as a result of our encounter with the work. These changes, in turn, alter our interpretation of the work, which in turn may lead to new changes in us, and so on. This wider circle, that involves the subject's changing anticipations is a main topic in the New Hermeneutics.

The "new" hermeneutics, Gadamer and Heidegger

The circle that involves the interpreting subject and the interpreted object is certainly the most intriguing, going to the core of how interpretation affects us and changes both subject and object. This is also the key issue between traditional and "new" hermeneutics. The new hermeneutics is usually associated with Gadamer and his teacher Martin Heidegger (1889–1976). However, its basic ideas go back to Edmund Husserl (1859–1938). Husserl's phenomenology is largely a study of the subjective perspective, the way in which all our experience, whether it be of the physical nature, of other human beings or of their actions and the products of these actions, such as texts, is imbued with meaning: there is always a web of anticipations involved, so that what we experience goes far beyond the patterns of irritations on our sensory surfaces. We are not aware of most of these anticipations. They form a horizon, a background, that for the most part is not thematized.

Husserl developed a special method for studying these anticipations, including the tacit ones that we are not normally aware of. This method, which he called the phenomenological reduction, is a special kind of reflection that makes it possible to bring our anticipations to consciousness and study their intricate structure. We can never uncover them completely, and we may make mistakes in recognizing them. Phenomenological reduction, like all other inquiries, is fallible and when we use it, we often discover that earlier findings have to be revised. This is partly because the reduction, like all other actions of ours, takes place within a horizon, which influences what we observe, but which largely remains unknown to us.

Husserl was especially interested in studying intersubjectivity, or what happens when we live in a society with others to whose anticipations we gradually come to adjust; the result is that our horizons become mutually attuned. This mutual adaptation mostly takes place silently and unnoticed, through common activities. However, it happens partly through actions and products of action that are intended for communication, such as speech or texts. When applied to speech or texts, the phenomenological analysis becomes meaning analysis of the kind one finds in traditional hermeneutics. However, since phenomenology extends the realm of meaning to all kinds of human experience, phenomenology becomes a kind of general meaning analysis, a generalized form of interpretation, or hermeneutics.

Heidegger therefore calls his version of phenomenology, which he took over from Husserl, "hermeneutics": "The phenomenology of Dasein [the human subject] is a hermeneutic in the primordial signification of this word, where it designates this business of interpreting."[7] Hermeneutics thereby becomes a means for getting insight into man's existence, which is the central theme in Heidegger's philosophy.

Gadamer, in *Truth and Method* and in several smaller works, applies this phenomenological conception of hermeneutics to the subject matter of traditional hermeneutics, the interpretation of texts. Husserl's notion of the horizon becomes particularly important in this enterprise. Gadamer emphasizes that when we read a text, our reading is shaped by anticipations we bring to our reading. Following Heidegger, Gadamer sometimes calls these anticipations fore-structures or fore-

---

[7]Martin Heidegger, *Sein und Zeit*. E. Husserl, ed., *Jahrbuch für Phänomenologie und phänomenologische Forschung*, Vol. VIII, 1927.

meanings. He also calls them prejudices (*Vorurteile*), and argues that we have inherited from the enlightenment a prejudice against prejudices. Rather than being something negative, prejudices in the sense of anticipations are unavoidable, writes Gadamer. However, one unfortunate drawback of this attempt to give an old word a new sense is that Gadamer's German word "*Vorurteil*," like its English equivalent "prejudice" has a strong negative flavor, and even more than its English equivalent has overtones of consciously made judgments. A most important feature of Husserl's notion of anticipation, or better fore-meaning or fore-structure, is that it is unconscious. We are not aware of it, and this is just why it is such a challenge to hermeneutics.

A main task of hermeneutics is to adapt our fore-meaning to the text. We must approach the text with openness, that is with awareness that we have fore-meanings and that the text may have a meaning that is incompatible with our fore-meaning. When we perceive a physical object we adapt our anticipations to the object: we revise those anticipations that do not fit until we reach an equilibrium. Similarly, when we encounter a text, we adapt our fore-meanings to the text: we revise our anticipations of what is expressed in the text until we find an interpretation that seems to us to be true or at least reasonable. That is, we adjust our interpretation and we adjust our opinions until we find that we can agree with the text. The criterion of understanding is this kind of "fusion of horizons."

Sometimes, in spite of our best efforts, we are not able to achieve this kind of total agreement. Then we have to aim for the second best: We attribute to the author a view which we regard as wrong, but we explain how he could plausibly have arrived at such a view, for example by pointing to his education and upbringing, his experience or other historical or psychological factors. This we will call "secondary understanding," contrasting it with total agreement, which we call "primary understanding." When we can achieve neither primary nor secondary understanding, we have to admit that we do not understand.

What is meaning? Quine and Davidson

The evolution in hermeneutics I have just traced has taken place on the continent, independent of and in isolation from similar developments in so-called "analytic philosophy."

There are philosophers, like Frege and Popper, who have held that in the study of linguistic meaning we are investigating a "third world," which is not of human making, but is akin to a Platonic realm, waiting for us to explore it. Frege's main argument for such a third world was that without it communication would be impossible. According to Frege, communication takes place when a speaker or writer expresses a certain meaning by help of a linguistic expression and a listener or reader connects the same meaning with this expression. A similar view often seems to have been taken for granted by hermeneuticists, however, apparently without awareness of the problematic philosophical presuppositions involved. An abstract world of meanings, without an account of how we get access to this world and why it is the same for all people, does nothing to explain language learning and communication. We owe it to Quine to have seen the emptiness of such an approach to meaning and communication.

The public nature of language

Philosophers and linguists have always said that language is a social institution. They have, however, immediately forgotten this and have adopted notions of meaning that are not publicly accessible and where it remains unclear how such entities are grasped by us.

Quine seems to have been the first to take the public nature of language seriously and explore its consequences for meaning and communication.[8] He begins with a situation where two people, each with their own language and view of the world, attempt to communicate. They have no previous translation manual to fall back on, no grammar or dictionary, but must carry out "radical translation," where they try to establish a grammar and a dictionary that they test out by observing one another's behavior.

Quine specifies two constraints that translation manuals have to satisfy. First, a condition on observation sentences, that is sentences which the other person assents to or dissents from only in certain observational circumstances. Such sentences should be translated into sentences that we assent to or dissent from in similar circumstances. Secondly, a principle of charity. Sentences which the other person accepts should not be translated into sentences which we regard as absurd, and sentences which the other person dissents from, should not be translated into sentences that we regard as banal.

As Quine points out, several different translation manuals can satisfy these constraints. Given that these two constraints are all the evidence there is for correct translation, Quine concludes that translation is indeterminate; there are several translation manuals between two languages, and they are all correct.

I shall not here discuss indeterminacy of translation, but I will concentrate on Quine's constraints on translation.

---

[8]W.V. Quine, *Word and Object*. Cambridge, Mass.: M.I.T. Press, 1960.

Problems with perception

I find the second of Quine's constraints, the principle of charity, well justified. It reflects an old and well-established hermeneutic principle and Quine supports it with good arguments. The first constraint, however, the observation constraint, both Davidson and I find very problematic. Not because observations are irrelevant to understanding and translation — their relevance will be a main theme of this paper — but because Quine defines observations in terms of the behaviorist notions of stimulus and response. Our problem is not the usual arguments against behaviorism, such as those of Chomsky against Skinner. Chomsky's arguments are pretty irrelevant against Quine's more discerning behaviorism.

Our problem is that we find that Quine through his focus on stimulus and response has forsaken the public nature of language. Stimuli can be *empirically* studied, but they are not *publicly* accessible. And according to Quine's fundamental insight, the emergence and development of language, the learning of language, and the use of language in communication must all be founded on publicly available evidence. In my daily life, where I learn and use language, I cannot observe the sensory stimuli of others. And I have never observed my own. How can I then compare the stimuli of others with those of my own, as Quine requires? The stimuli are encumbered by the same problem as Frege's "Sinne," they are not publicly accessible. What the child learns to associate with words, are neither "Sinne" nor stimuli, but things in the surrounding world.

Quine developed his view on stimuli further in *Ontological Relativity* (1969)[9] and ended the book with the open problem of "saying in general what it means for two subjects to get the same stimulation or, failing that, what it means for two subjects to get more nearly the same stimulation than two others". In a lecture series in Oxford in 1974 where Quine and Davidson also participated, I criticized Quine's emphasis on stimulations thus:

> In criticism of this proposal of Quine's, I would like to suggest that identifying stimulations with triggering of nerve endings sets us off on a wrong track. By talking about the triggering of the sensory receptors we are already going too deeply inside the skin. Language being a social phenomenon, the basis for language learning and communication should also be publicly accessible without the aid of neurophysiology. This is a point repeatedly emphasized by Quine himself. In fact, on page 157 of *Ontological Relativity* he says that homology of receptors 'ought not to

---

[9] Quine, *Ontological Relativity and Other Essays*. New York: Columbia University Press, 1969.

matter'.¹⁰

Also, already in the very opening sentences of *Word and Object* Quine stressed how language learning builds on distal objects, the objects that we perceive and talk about:

> Each of us learns his language from other people, through the observable mouthing of words under conspicuously intersubjective circumstances. Linguistically, and hence conceptually, the things in sharpest focus are the things that are public enough to be talked of publicly, common and conspicuous enough to be talked of often, and near enough to sense to be quickly identified and learned by name; it is to these that words apply first and foremost.¹¹

Why, then, did Quine turn to stimuli? He saw, I think, clearer than it had ever been seen before, how intricate the notion of an object is. We cannot determine through observation which objects other people perceive; what others perceive is dependent upon how they conceive of the world and structure it, and that is just what we are trying to find out. When we study communication and understanding, we should not uncritically assume that the other shares our conception of the world and our ontology. If we do, we will not discover how we understand other people, and we will not notice the important phenomena of indeterminacy of translation and of reference. Already in chapter 3 of *Word and Object*, the chapter following the chapter where he introduces stimuli, Quine discusses the ontogenesis of reference, and the discussion of this topic takes up several of the following chapters.

The early Davidson: "Maximize agreement"

Davidson's theory of radical interpretation is similar to, but also interestingly different from Quine's theory of radical translation. One highly important contribution of Davidson to the whole theory of understanding, understanding language as well as understanding action and the mind, is his idea of comparing the problem of separating belief from meaning in the understanding of language to the problem of separating belief from value — or desire, or generally, pro-attitude — in the explanation of action. Since one component in each pair, that of belief, is the same, Davidson is able to add observation of actions to our evidential basis for interpretation and translating of language.

---

[10] "Meaning and experience." In Samuel Guttenplan, ed., *Mind and Language: Wolfson College Lectures 1974*. Oxford: Oxford University Press, 1975, pp. 34-35.

[11] *Word and Object*, p. 1.

I will, however, here concentrate on that part of Davidson's theory of interpretation that has a counterpart in Quine's theory of translation. In order to compare the two theories let us transform Davidson's theory into a theory of the conditions a correlation between two languages must satisfy in order to be a translation. There are various reasons why Davidson prefers a theory of interpretation to a theory of translation manuals, but they are not pertinent to my aim in this paper.[12]

I shall argue that there is an early and a late version of Davidson's theory. I will here concentrate on the early theory, that is most pertinent to Gadamer. Davidson held his early theory up to 1973. This early theory differs from Quine's on the following two points:

1. Davidson replaces Quine's systematization via grammar with a systematization by means of Tarski's theory of truth. This change reflects the fact that the systematization concerns semantics: one wants to see how the semantic features of complex expressions depend upon the semantic features of their component expressions. More accurately: given that one knows, through behavioral evidence, which sentences a person assents to — that is, regards as true — and which sentences he dissents from — regards as false — we try to segment these sentences into recurrent parts, that is words, and to find extensions and references for these words that make most of the sentences that the person assents to true and most of the ones he dissents from false. This idea of interpreting the other's sentences so as to make the sentences he assents to true, is a point of similarity between Davidson and Gadamer, although Gadamer does not mention Tarski.

This proposal by Davidson could be looked upon as applying Tarski upside down. While Tarski assumed that we know the extensions and references of the smallest components and built up from there, Davidson starts with the truth and falsehood of sentences and tries to determine the parts and their semantic features from there.[13]

I regard this first proposal of Davidson's as an improvement upon Quine. And Quine has accepted it.

2. Davidson's second proposal is to fuse Quine's two constraints on translation that I outlined above into one single constraint, a sweeping principle of charity that he expresses as a maxim: *maximize agreement.* That is: try to correlate the two languages in such a way that the sentences to which the other person assents are correlated with sentences to which we assent, and sentences from

---

[12]See Donald Davidson, "Radical Interpretation." *Dialectica* 27 (1973), 313-28, esp. pp 316-17.
[13]See, e.g., Davidson, "Belief and the Basis of Meaning," *Synthese* 27 (1974), 318.

which he dissents are correlated with sentences from which we dissent.[14]

This simple constraint was the only condition Davidson put on translation in his early writings. He had recognized Quine's problems in connection with perception, and he formulates his constraint without any appeal to perception. In Davidson's early writings there is no mention of perception as one of the factors one has to take into account when one interprets somebody else.

It would certainly simplify matters if perception did not have to enter the picture. However, I found Davidson's maxim problematic, and in 1973, on a walk in the hills near Brügg, Switzerland (the town where Dale and Tina Jacquette came to live many years later and where Dale died 8 days after their 40th wedding anniversary). I discussed with Davidson the indispensability of perception for interpretation. I used the following example of a rabbit behind a tree, and was curious as to how he would handle it:

I am together with a person who speaks a language which I do not know, but would like to learn. He frequently uses the phrase 'Gavagai' and I have formed an hypothesis that it has to do with rabbits. While we are in a forest and I note a rabbit I try out the phrase 'Gavagai'. However, my friend dissents. According to Davidson's maxim of maximizing agreement this would be a reason against my hypothesis that 'Gavagai' should be translated 'Rabbit'. If I now discover that there is a big tree between my friend and the rabbit, I immediately have an explanation for our disagreement: I take it for granted that my friend, like me, is not able to see through trees and that he therefore does not think that there is a rabbit there. I even take my friend's dissent as confirming my hypothesis; I do not expect him to believe that there is a rabbit there.

The maxim of maximizing agreement hence has to be modified into *maximize agreement where you expect to find agreement*. Here both of Quine's constraints on translation come in, the observational and the principle of charity. Interpretation recapitulates epistemology, and Quine's two principles reflect the two main ingredients in epistemology: perception and reason.

The rabbit-behind-the-tree example illustrates how the perceptual situation which we assume the other to be in may be decisive for the beliefs we ascribe to him and thereby for how we interpret and translate what he says. When Davidson was

---

[14]The expression "maximize agreement" recurs in many of Davidson's papers from this period, for example in "Truth and Meaning" (1967), where it is explained as follows: "The linguist will then attempt to construct a characterization of truth-for-the-alien which yields, so far as possible, a mapping of sentences held true (or false) by the alien on to sentences held true by the linguist." (Page 27 of the reprint in Davidson's *Inquiries into Truth and Interpretation*. Oxford: Oxford University Press, 1984).

confronted with this example, he agreed that perception is important for translation and interpretation. In his later writings Davidson gives prominence to perception.

To sum up: According to Quine, communication and understanding are based on our observation of one another's behavior, not only in the sense that the behavior provides evidence upon which we can base our judgments concerning meaning, but in the much more radical sense that meaning is a product of this publicly accessible evidence. That is, meaning, unlike nature, was not there before public interaction began, waiting to be discovered, but it has been produced, and is continually being produced, through this interaction. This production is a co-operative enterprise, where previous generations and specialists on various subjects (theoreticians and practitioners) have done their part, and created the semantic construction that the learner of the language strives to master. The production process still goes on, through our introduction of new expressions and through our using old expressions in new ways, as for example in metaphors. Where the evidence leaves off, there is nothing more to be right or wrong about. This is the gist of Quine's thesis of "indeterminacy of translation."

For Quine, epistemology and meaning are intertwined: one of our main tasks when we try to understand a text or another person is that we try to interpret the other in such a way that he comes to have views that it would be reasonable for this person to have, given the person's background and past and present experiences. That is, when we translate what another person says or writes into our own language or idiom, we seek to translate it in such a way that we come to attribute to him beliefs that we would expect him to have, given our theory of how people acquire and alter their beliefs: "the more absurd or exotic the beliefs imputed to a people, the more suspicious we are entitled to be of the translations."[15] Though the way of arguing for it is very different, this is also essentially Gadamer's view.

To conclude: There is a striking similarity between Gadamer, Quine and Davidson in their emphasis on how truth and agreement is crucial for interpretation. Where they differ, is that Quine and Davidson find the notion of meaning deeply problematic. Given that meaning is the central notion in hermeneutics, hermeneutics cannot ignore the radical exploration into the nature of meaning that Quine and Davidson have begun.

---

[15] Quine, *Word and Object* (1960, 2013), p. 69.

CONTENT AND OBJECT IN BRENTANO

GUILLAUME FRÉCHETTE
University of Salzburg, Austria.
`guillaume.frechette@sbg.ac.at`

1 The received view

It has usually been maintained that Brentano's theory of intentionality never actually distinguished between the content of an act and its object, and that the distinction was introduced by Meinong and Höfler (1890), then more systematically by Twardowski (1894), and later by Husserl (1900/1). Taking the *Psychology from an Empirical Standpoint* as a reference point, it may indeed seem that Brentano's theory of intentionality offers little room for a distinction between content and object. Intentionality is introduced there as a mark of mental phenomena, in contrast with physical phenomena. The latter are signs of an outer reality. They are not physical in the proper sense of the word: rather, they are the content of mental phenomena and they lack intentionality. In this context, a distinction between content and object seems superfluous, since the mark of the mental is determined without any reference to the objects of outer reality. This is the basic idea behind the "non-distinction" view.

For this reason, many readers of Brentano considered his "intentional inexistence" of the presented objects to be some sort of "diminished existence." The non-distinction view was defended from early on by the editors of Brentano's Nachlass — Oskar Kraus, Alfred Kastil, and Franziska Mayer-Hillebrand — to such an extent that they adapted many manuscripts to fit the non-distinction view.[1] Furthermore, this editorial policy based on the non-distinction view was clearly the source of Roderick Chisholm's influential interpretation of Brentano's intentional

---

[1] See for instance Brentano (1959), where the editor Mayer-Hillebrand changed unilaterally all instances of *Inhalt* (content) into *Gegenstand* (object) in her edition of the 1885/86 lectures on psychology and aesthetics, thereby obliterating the linguistic (and ontological) distinction originally used by Brentano. Her motivation for doing so was to propose an edition complying with Brentano's alleged later views (Brentano 1959, 234). Unfortunately, this motivation is not only historically and editorially questionable, but it is also not clear that Brentano's later views really imply the non-distinction view.

objects as mental contents whose mode of being is "short of actuality but more than nothingness" (Chisholm 1967, 201),[2] as well as of Barry Smith's interpretation of intentional objects as having a "diminished form of existence ... 'in the mind' " (Smith 1995, 44), and Dale Jacquette's notion of "immanent intentionality,"[3] to name only a few of these interpretations based on the non-distinction view.

Recent research on Brentano's lecture manuscripts from the 1870s and 1880s, however, has shown that Brentano discussed the distinction between content and object at length in the very lectures that were attended by Meinong, Höfler, and Twardowski.[4] Attributing to Brentano the non-distinction view of content and object of presentations (as has often been done), and the corresponding thesis that intentionality is a relation with an internal, mental entity no longer seems to be historically accurate; it also imposes strict limitations on the intentionality thesis which are not even necessitated by Brentano's own positions. One of these limitations is the so-called "methodological phenomenalism" sometimes attributed to Brentano (see Simons 1995, xvii; Crane 2006), according to which science, and philosophical investigations as well, can only study phenomena; this view can only hold if one also maintains the non-distinction view. The non-distinction view also makes Brentano's philosophy quite unattractive for any theory of meaning based on the distinction between sense and reference. These limitations on Brentano's concept of intentionality are particularly difficult to maintain when one considers his lectures on logic from the late 1860s and early 1870s, in which he clearly states and develops the distinction between content and object; moreover, his lecture notes on descriptive psychology from the mid- and late-1880s also basically follow the same concern, as did his logic lecture notes from the Vienna period.[5] One finds in these documents an explicit concern with the distinction itself and its application in a more general theory of intentionality.

I will discuss these lectures and the quotes themselves in section 3. Before that, in section 2, I would like to suggest that Brentano's own conception of philosophy speaks in favour of a more general reading of the intentionality thesis than the one suggested by Dale Jacquette's "immanent intentionality" and by the sympathizers of the Chisholmian reconstruction of Brentano.

---

[2] See also Chisholm (1960, 4-5); (Chisholm 1957, 169).
[3] See Jacquette (1990, 178); (2004, 105ff.).
[4] This is for instance the case with the numerous lectures delivered by Brentano in Vienna between 1874 and 1891, most notably on logic, descriptive psychology, and ethics.
[5] Some of this material will be published soon in Brentano (forthcoming).

2 Brentano's world-view

It is tempting to argue that Brentano was not a systematic philosopher. After all, he saw the idealistic system-builders of the nineteenth century as the proponents of a kind of thinking representative of philosophy in its deepest state of decline. In contrast with idealist systematic philosophy, Brentano's philosophy, at least in its published form, seems much more fragmented, often dealing with particular problems and rarely with overarching considerations. This is the case with his philosophy of perception, for instance, in which he devotes considerable effort to discussing the properties of colours and sounds.[6] In many other places too, the concern with establishing the conditions of the possibility of knowledge *in general*, characteristic of neo-Kantian philosophy, is completely absent.

Brentano was firmly opposed to the neo-Kantian separation between the natural and the human sciences on the basis of their supposedly distinct methodologies. But this cannot be taken to mean that he was not preoccupied by questions of structure and methodology, and other questions pertaining to what is usually called the systematicity of philosophy. This is most clearly shown by his awareness of the advances in the empirical sciences, especially in physiology and psychology but also in biology and physics. Given the increasing specialization of these disciplines, some philosophers, inspired by Kant's critique, would prefer to isolate the methods of philosophy from the rest of the sciences. Brentano's attitude however was quite different: as is well known, the fourth of his twenty-four *Habilitationsthese* of 1866 states that the methods of philosophy are *identical* to those of natural sciences. The thesis doesn't mean that philosophy should simply subordinate itself to natural sciences, for example by simply endorsing some hypotheses from the natural sciences and taking them for granted as a basis for philosophical thought — this strategy has sometimes been used in twentieth-century philosophy of mind. Rather, it means that if philosophy is to retain its credibility as an endeavour concerning the "big picture" — that is, one that addresses the most fundamental questions — it cannot accept being cut loose from the empirical sciences.

With this in the background, Brentano rejected philosophical systems as these were understood classically, that is, as systems of truth ranging over all domains of philosophy and obtained either by dogmatic construction or by transcendental deduction, both of which he considered symptoms of philosophical decay. Rather, he was turning to an older conception of philosophy inspired by Aristotle, following the idea that philosophical inquiry is by nature a *Weltanschauung*, a "world view," which constitutes a whole in which truths of different domains are structured and

---

[6]Brentano (1907).

dependent parts. Toward the end of his life, he summarized his own understanding of Aristotle in *Aristotle and His World View*, which gives insight as much on Brentano's own understanding of Aristotle as on Brentano's own worldview. In a review of the book, his student Hugo Bergman aptly remarked that for Brentano, the path through wisdom (*Weisheit*) must start with the knowledge of individuals, and proceed from there through metaphysics, to end up with a general worldview desirable for humanity (Bergman 1912).

With this notion of *Weltanschauung* or "world view", it seems that Brentano found a compromise between, on the one hand, an overly fragmented approach to philosophy in which individual questions are treated completely independently with no perspective on the "big picture," and, on the other hand, a dogmatically systematic approach in which attempts to protect the internal coherence of the system lead to increasing insulation from reality. Such a world view aims to integrate truths obtained from the empirical sciences, philosophical analysis, and psychological descriptions in a unified way that allows for adjustments in the course of the development of the individual disciplines.

In short, although he was not an advocate of systematic philosophy in the sense of a dogmatic system of philosophy, Brentano was a systematic philosopher to the extent that he wanted to integrate into his *Weltanschauung* (or the *Ganze* of his view) the truths obtained from the empirical sciences, philosophical analysis, and psychological description, as mereological parts of the whole.[7] The fact that adjustments between the parts may remain open in the course of the development of the individual disciplines, in a loose analogy with the parts of a biological organism during its actual development, is reminiscent of the organic mereology of Friedrich Trendelenburg, the Prussian author of the *Logische Untersuchungen*, who was also Brentano's *Lehrer* in Berlin.

With this idea of a world view in mind — the idea that the task of philosophy is to seek, starting from the knowledge of individuals, and coming back there through metaphysics, a general world-view desirable for humanity — we may ask ourselves if the view generally and traditionally attributed to Brentano is actually an exact and faithful rendering of a more general world view that is Brentano's own. Is the view that intentionality is the property of mental phenomena according to which they "contain in themselves" a so-called physical phenomenon, and that the objects of intentional acts are simply mental objects with a "diminished existence" really a building block of Brentano's world view? In other words, if intentionality is supposed to be *essentially* a way of distinguishing between two kinds of mental appearance

---

[7]On a different but compatible take on Brentano as a systematic philosopher, see Kriegel (forthcoming).

(physical and psychical), and if it is supposed to be a central aspect of Brentano's world view and not some marginal consideration, then at some point we will have to consider some variety of phenomenalism as constitutive of this world view. My point here is simply this: it is hardly disputable that intentionality is one of the main features of Brentano's psychology. In fact, we may well say that his whole descriptive psychology presupposes the concept of an intentional relation. His metaphysics too refers to intentionality — that is, to the fact that some entities are *entia objectiva* — in many significant places. Generally, however, if his philosophical world view is supposed to integrate the truths of the empirical sciences with philosophical analysis and psychological description, why would it limit such a multifaceted device as intentionality to a relation that holds strictly between mental entities, without any kind of mediating term? How is the theory of intentionality to fit with the truths of the empirical sciences, or with the development of language and communication?

One possible answer to these questions is simply to see Brentano's world view as endorsing this limited account of intentionality as a relation between mental entities, and holding the view that in their accounts of reality both philosophy and the natural sciences are limited to mere phenomena. This reading of intentionality has often been endorsed (Simons 1995; Smith 1995; Chisholm 1967), and trivially it is absolutely true as a description of the project of an empirical psychology presented by Brentano in 1874. It is also correct to describe the 1874 empirical psychology project as some kind of "methodological phenomenalism," as suggested by Peter Simons, where "methodological" means a reduction of the scope of Brentano's phenomenalism in order to exclude his metaphysics from its scope.

Under this description — that is, intentionality as a relation between two mental entities — which lays the ground for Brentano's alleged methodological phenomenalism, there is no place for a distinction between the content and the object *of an act*, and it makes perfect sense *not* to introduce such a distinction as a philosophical interpretation when it is not needed or introduced.

But still, given the idea that Brentano had this project of integrating into one single *Weltanschauung* (a mereological whole structured in an organic way) the truths obtained from the empirical sciences with the products of philosophical analysis and psychological description, the question remains: is his concept of intentionality meant to fit only his early, unfinished project of an empirical psychology? (In which case its place in the general *Weltanschauung* would be relatively restricted, or would be characteristic of a phenomenalistic *Weltanschauung)* Or was the concept rather meant in a more general sense, which would include its introduction in the context of the project of empirical psychology, but would also have further domains of application?

In the next section (section 3), I will provide some evidence for the thesis that the concept was indeed meant in a more general sense. In the final section (section 4), I will come back to the reconstruction of Brentano's position as a methodological phenomenalism in order to show that such a reconstruction is not compatible with the world view defended by Brentano.

3 The distinction

As pointed out earlier, it has often been maintained that the distinction between content and object was first introduced by Brentano's students. Twardowski is often credited with this distinction in his habilitation thesis:

> In comparing the act of presenting with painting, the content with the picture, and the object with the subject matter which is put on canvas — for example, a landscape — we have also more or less approximated the relationship between the act on the one hand and the content and the object of the presentation on the other. ...We shall say of the content that it is thought, presented, *in* the presentation; we shall say of the object that it is presented *through (durch)* the content of the presentation (Twardowski 1894/1977, 15 — 16).[8]

It was this distinction, said Meinong in 1899, that paved the way for a distinction between content and object. Now, in 1899 Meinong was in a very difficult personal and professional relationship with Brentano, to such an extent that it is perfectly understandable, knowing now the full background of his conflict with Brentano, that he preferred to quote a *published* work by a student of Brentano over an *unpublished manuscript* by Brentano himself.[9] But Twardowski can only be credited here at most with having given the first published form to the distinction, but not with the distinction itself.[10] Brentano's early logic lectures from the late 1860s and early 1870s already present this distinction quite explicitly:

---

[8]"Indem wir den Vorstellungsact mit dem Malen, den Inhalt mit dem Bild und den Gegenstand mit dem auf der Leinwand fixierten Sujet, etwa einer Landschaft, verglichen haben, ist auch das Verhältnis, in welchem der Act zum Inhalt und Gegenstand der Vorstellung steht, annähernd zum Ausdruck gelangt. ...Vom Inhalt werden wir sagen, dass er *in* der Vorstellung gedacht, vorgestellt werde; vom Gegenstand werden wir sagen, dass er durch den Vorstellungsinhalt (oder die Vorstellung) vorgestellt werde. Was *in* einer Vorstellung vorgestellt wird, ist ihr Inhalt; was *durch* eine Vorstellung vorgestellt wird, ist ihr Gegenstand. (Twardowski 1894, 17-18).

[9]On the conflict between Meinong and Brentano, see Marek (forthcoming).

[10]In fact, Twardowski attended to the three lectures on descriptive psychology delivered by Brentano in Vienna in 1887/88, 1888/89, and 1890/91. Near the end of the first lecture (see Brentano (forthcoming)), Brentano explicitly formulates the distinction between content and object of presentations.

> The name manifests a mental phenomenon, it means the content of a presentation as such (the immanent object?), it names that which is presented through the content of a presentation. Of this we say that the name is attributed to it. We name any real objects of presentation the external object of the presentation (when it exists). (One names through the mediation of meaning).[11]

We have here a quite precise identification of the concept of an "immanent object" with the concept of a "content of presentation." When using names, suggests Brentano, presenting as an activity seems to involve two directions, so to speak: when I want to convey to you that you should look at the moon, I might simply say to you, in the appropriate context, "The moon!", after which you normally would look up at the sky searching for the moon. When you hear "The moon!", you have a content of presentation, the content of your thought of the moon at this very moment, which is that *through which* you later look up in order to find the object in the sky. Twardowski's distinction of presented-in (for contents) and presented-through (for objects) is nothing new: it had been suggested in exactly these terms by Brentano more than twenty-five years earlier.

Two further important points are contained in the last quote. First, that which one names is a real object, which may or may not exist. Only real existing objects are external objects of presentations. Second, the content of a presentation serves as a *Bedeutung* (a *Sinn* in Fregean terminology) in acts of meaning. To the first point: Brentano distinguishes at many places between reality and existence. This distinction can be spelled out in the following way: unicorns are real objects, but obviously are non-existent; Hanoverian horses are real objects too, but they enjoy over unicorns the formidable advantage of existing, and only this advantage can give

---

[11]"Der Name gibt kund ein psychisches Phänomen, bedeutet den Inhalt einer Vorstellung als solchen (den immanenten Gegenstand?), nennt das, was *durch* den Inhalt einer Vorstellung vorgestellt wird. Davon sagen wir: es kommt ihm der Namen zu. Man nennt die etwaigen wirklichen Gegenstände der Vorstellung was, wenn es existiert, äußerer Gegenstand der Vorstellung ist. (Man nennt unter Vermittlung der *Bedeutung*)." EL81, p. 13528 (around 1867–1872). We find an alternative formulation in the logic lectures from the early 1870s: The name designates in a certain way the *content* of a presentation as such, the immanent object: in a certain way that *which is presented through the content of a presentation*. The former is the meaning of the name. The latter is that which the name names. Of this we say that the name is attributed to it. It is that which, if if exists, is the external object of the presentation. One names through the mediation of meaning (EL80, 13018). German original: "Der Name bezeichnet in gewisser Weise den *Inhalt* einer Vorstellung als solcher, den immanenten Gegenstand. In gewisser Weise *das, was durch den Inhalt einer Vorstellung vorgestellt wird. Der erste* ist die *Bedeutung* des Namens. Das *zweite* ist das, was der Name nennt. Von ihm sagen wir, es komme der Name ihm zu. Es ist das, was, wenn es existiert, äußerer Gegenstand der Vorstellung ist. Man nennt unter Vermittlung der *Bedeutung*."

them the privilege of filling the object position in the act-content-object schema. To the second point: in acts of meaning, the distinction between content and object follows from the distinction between two functions of the expression: to express a mental content and to refer, on the basis of the first function, to an object.

Now it seems that we can distinguish at least between two different understandings of this second point: following the first reading of the distinction between meaning a content and presenting an object *through* the content, one could say the distinction between content and object of presentations appears only at the expressive level. In other words, if no contents are expressed, the distinction does not come into existence. Following the second reading, one could rather say that the distinction between content and object, as it appears on an expressive level, is simply one particularly good way to illustrate a distinction that is otherwise at play in all acts (expressive or not) of presentation. At this point, I leave the question open as to which reading is the more appropriate. In any case, this distinction is spelled out in another way a few pages later:

> The content of presentation is named, [i.e.] the objects-presented-as-they-are-presented (*als das, als was sie vorgestellt werden*). The content of judgment is stated, [i.e.] truths or falsities of the objects-presented-as they-are-presented. The content of feeling is evoked (*ausgerufen*), [i.e.] the goodness or badness of *A* as *A* of the objects-presented-as-they-are-presented. [*in the margin:*] Man is a species = the content of the presentation of a man is a *species* = the *meaning* of the word man is a *species*.[12]

The idea here seems to be the following: when I say "the highest bell tower in Salzburg" to express the content of my presentation, this also *names* an object through the mediation of the content, which is in this case what Brentano calls the "object-presented-as-it-is-presented." Here, the hyphens have a double purpose:

---

[12] "Genannt wird der Inhalt von Vorstellungen : das sind die Gegenstände, als das, als was sie vorgestellt werden. Ausgesagt wird der Inhalt von Urtheilen: das sind die Wahrheiten oder Falschheiten der Gegenstände als das als was sie vorgestellt werden. Ausgerufen wird der Inhalt von Gefühlen: d.i. die Güte, Schlechte von A als A der Gegenstände als das, als was sie vorgestellt werden." EL 81, 13609 (around 1867–1872). "Der Name gibt kund ein psychisches Phänomen; bedeutet den Inhalt einer Vorstellung als solches (dem immanenten Gegenstand?), nennt das, was *durch* den Inhalt einer Vorstellung vorgestellt wird. Davon sagen wir: es komme ihm der Namen zu. Man nennt die etwaigen Gegenstände der Vorstellung das, was, wenn es existiert, äußerer Gegenstand der Vorstellung ist. Man nennt unter Vermittlung der *Bedeutung*. [Am Rand:] Mensch ist eine Species = der Inhalt der Vorstellung eines Menschen ist eine Species = die *Bedeutung* des Wortes Mensch ist eine *Species*" (1870/1 (WS) od. 1873 (SS)). Fragmente aus der Würzburger Logik, EL81, 13528.

they are meant to emphasize that the expression used is not further decomposable, but they also serve to signal that an expressive function (naming, stating, evoking, etc.) has been used to refer to the object. It seems therefore that the distinction between the content and the object of presentations appears for Brentano when both the expressive and referential functions of language are at play. By stating "Snow is white," I want to express a judgment I now have, so that you will believe that I think that snow is white; but by doing this, I also want you to look at that portion of the world that my judgment is about, in order for you to understand what my utterance of "Snow is white" refers to. Something has a referential function on the basis of the realization of its expressive function.

Take however a sensory experience, such as seeing blue, feeling pain, or blindly acknowledging the existence of the blue seen (which are often called by Brentano "intuitions", "blind intuitions" or "blind acknowledgements") which remains unexpressed. In this case, the problem of the two readings of the quotation raised above clearly appears: on the strong reading of the distinction between content and object (where the distinction only comes into existence when expressive acts are involved), we would have to say that my seeing blue cannot allow for a distinction between the content of the seeing and its object. On a phenomenological basis, the strong reading seems reasonable: it would seem odd phenomenologically, in such an experience, if unexpressed presentations and unvoiced judgments gave us the object-presented-as-it-is-presented in contrast with the object *tout court* (the blue, the pain). It would seem more natural to suppose that the phenomenology of such an experience gives us a perceptual content *tout court*, as purely indistinguishable from the object of the seeing, the feeling, or the blind judging. Following the strong reading seems plausible in this case, but it cannot account for a general distinction between content and objects of all mental acts: it can only provide a local distinction along these lines: As soon as the corresponding mental act is expressed intentionally, that is, as soon as it is made public, a distinction between content and object seems to be introduced automatically.

In the background of our analysis, another plausible (and perhaps slightly more precise) way of rephrasing this would be to say, "When I express the content of a presentation and when the expressive and referential functions of language are realized, something is named: the the content (or the object-presented-as-it-is-presented) serves as a mediator for the referential function of the expression, and the object may or may not exist."

This characterization of the actual content of the presentations (the objects-presented-as-they-are-presented: *Gegenstände, als das, als was sie vorgestellt werden*), their distinctness from the objects of presentations as objects tout court, their nature as the referents of the names, and the semantic setting in which this distinction is introduced prefigure Frege's distinction between sense (*Sinn*) and reference (*Bedeutung*). Even if Brentano, in contrast with Frege, seems not to consider his contents to be anything more than individual mental entities, he seems to attribute to them not simply a presentational nature, but also, and more importantly, a *representational* one. This becomes clear when one takes into consideration the marginal note added by Brentano at the end of the passage quoted above. There he is obviously referring to the medieval understanding of species as representational and mediating entities. When we say that man is a species, we mean by this that the content of the presentation of a man is a *species*. While this content is basically mental (in mente), it gets its generality from abstraction. This general content, so to speak (the content as a species) is the *Bedeutung* of the word "man." In EL80, another manuscript from the lectures on logic, Brentano stresses this point in a different way:

> Against [the idea] that the content of a presentation is the meaning, it has been objected [...] [that] when I say: the sun rises, I do not mean that the content of my presentation rises; I speak of an external process. Answer: for this, we only need the external object to be what is named: it does not need to be the meaning. Rather, meaning is the content of the presentation "sun": the object is named through its mediation. Against [the idea] that objects are designated, it has been said: 1. Often, an object is missing. In that case, names would not mean anything. Answer: 1* They designate the objects but they do not mean them: rather, they name them. Therefore, the word is not without meaning. (EL 80) [13]

Quite unmistakably, Brentano takes the content of a presentation to be the *Bedeutung* (the *Sinn* in Fregean terminology), the content as a species, and this *Bedeutung* is clearly distinct from the object, which sometimes may be missing. I have focused here on earlier texts from the late 1860s and early 1870s. But we find similar considerations about the distinction between content and object in later manuscripts as well, for example, in the preparatory notes to his lectures on descriptive psychology from 1885 — 1890:

---

[13]"Dagegen, dass der Inhalt der Vorstellung die *Bedeutung* sei, wurde eingewandt [...]: Wenn ich sage, die Sonne geht auf, so meine ich nicht, der Inhalt meiner Vorstellung geht auf, ich spreche von einem äußeren Vorgang. Antwort: Dazu genügt, dass das äußere Objekt das Genannte ist, die *Bedeutung* muss es deshalb nicht sein, diese ist vielmehr der Inhalt der Vorstellung Sonne, unter deren Vermittlung das Objekt genannt wird. Dagegen, dass die Gegenstände bezeichnet werden, wurde gesagt: 1. Es fehle oft ein Gegenstand. Also würden die Namen nichts bedeuten. *Antwort*: 1'. Sie bezeichnen wohl die Gegenstände, aber bedeuten sie nicht, sondern nennen sie. Das Wort ist also nicht ohne *Bedeutung*." EL80.

> The presented does not need to be simply because it is presented. To be and to be presented are different things. Only as presented must [the presented] be, but not as object-presented-as-it-is-presented.
>
> For example, I am presenting the goddess Venus. In this case, what I am presenting does not exist. But a presented-Venus exists thanks to the fact that I present Venus. I call the presented as presented [the presented-Venus, GF] the content of a presentation. I call the object-presented-as-it-is-presented the object of the presentation. When something is presented, there always is a content. But there is often lacking an object of the presentation. For one single content of presentation there might be many different objects. And for one single object there might be many different contents. (PS48/1885 — 1890)[14]

Here, Brentano uses the earlier terminology of "objects-presented-as-they-are-presented" (*als das, als was sie vorgestellt werden*) differently. In the logic lectures, this expression was meant to characterize the content of a presentation, while in the lectures on descriptive psychology, it is obviously used as a characterization of the object of a presentation. This fluctuation is bothersome, but it does not challenge the distinction. What is relevant in the last quotation could be expressed in the following way: while Brentano spoke in the late 1860s and early 1870s of a distinction between content and object on the level of expressed presentations, he presents this distinction in the lectures on descriptive psychology in terms of a difference between two kinds of objects (the presented-Venus *vs* Venus *tout court*). Here, when I present Venus without expressing anything, the presented-Venus is the content of my presentation (some immanent object), while there is no object *tout court* — and if there were one, it would be a real object. Understanding the distinction in ontological terms (and not merely on the basis of expressed contents) might capture the problem of intuitive contents described above in a more satisfying way: when I have a perceptual experience of blue without expressing the content of my presentation, there is a perceptual content (there is an immanent object), and there is (or isn't) an object *tout court* towards which I am directed through the perceptual content.

---

[14]"Das Vorgestellte braucht deshalb, weil es vorgestellt wird, nicht zu sein. Es ist etwas Anderes, sein und vorgestellt sein. Nur als vorgestellt muß es sein, nicht aber als das als was es vorgestellt wird. z.B. Ich stellt mir die Göttin Venus vor. Das, was ich vorstelle, existiert in dem Falle nicht. Aber eine vorgestellte Venus existiert aber dadurch, daß ich die Venus vorstelle. Das Vorgestellte als vorgestelltes nenne ich Inhalt der Vorstellung. Das Vorgestellte als das als was es vorgestellt wird, wenn es ist, Gegenstand der Vorstellung. Wenn etwas vorgestellt wird, so ist immer ein Inhalt. Aber es fehlt oft ein Gegenstand der Vorstellung. Es können einem Vorstellungsinhalt viele und verschiedene Gegenstände entsprechen. Es kann auch ein Gegenstand vielen verschiedenen Inhalten entsprechen."

This way of spelling out the distinction proposed in the last quotation would speak in favour of the weak reading of the distinction between content and object.

To put it slightly more simply, and keeping the same use of hyphens as introduced earlier, we could say that the presented-chair, like the presented-unicorn, the presented-Aphrodite, or the *Vorgestellte*, as Brentano says, is not subject to existence at all. The presented-Aphrodite, through the simple fact that I have a presentation of Aphrodite, is brought in some sense into existence (though in a misleading sense of "existence"), but this means simply that in that case some content presenting Aphrodite is present to my consciousness, nothing more. What is presented through this content of presentation is the object, if it exists: the presented-chair has an object but the presented-Aphrodite does not.

Again, it seems that the weak reading of the distinction between content and object misses the idea that this difference seems to be phenomenologically accessible only in acts which are linguistically expressed or expressible. The last quotation cannot completely dismiss the strong reading of the distinction. Looking back at an experience with no intrinsic linguistic feature (the Brentanian "intuition"), such as seeing blue, I cannot really say whether the seeing actually involved (or not) a presented-blue *and*, incidentally, the object *tout court*. The reason why I cannot say so does not seem to be a matter of not paying enough attention to the phenomenology of seeing. It just seems to be a constitutive feature of such experiences with no intrinsic linguistic features that the question of the fittingness of the perceived with reality just cannot be raised. When I feel for the first time the peculiar pain in the lungs caused by pneumonia, without any knowledge or plausible hints at hand as to how to interpret this pain, I may well believe that the pain is caused by a muscle cramp, believing thereby the object of my pain to be a muscle rather than a lung. After being informed by a radiologist that the feeling described applies to a lung and not to a muscle, it would not be quite plausible to explain this by saying that I was misled by my phenomenological experience: that I *should* have seen the difference by paying more attention to the experience, etc. The issue of the distinction between content and object — when, why, and how this distinction emerges in experience — is obviously not to be solved by forcing conceptual distinctions into all kinds of phenomenological experiences (in order to assess thereby their universality), but simply by recognizing it when it actually emerges. Following Brentano's ideas as quoted here, I would suggest that this distinction emerges only on the basis of a content of presentation which either has intrinsic linguistic features (e.g. a conceptual presentation) or is linguistically expressed.

Finally, let me mention a last passage, this one from the Vienna lectures on logic at the end of the 1870s, where all the elements already highlighted are mentioned, but where another important one is added to the list:

> Names are expressions of presentations. But the presentation is not that of which we say that the name names it. The name rather expresses the presentation in such a way that it names that which is presented by the presentation, and it names it under its mediation and for this reason completely or incompletely determined (or indetermined) in the same way as it presents it. In this way, the presentation is the sense (*Sinn*) of the name; the thing is that which is named by the name and in the most proper sense that which is designated through it. ... Naming is not a manifesting of the named, but only an awakening of a presentation (*ein in die Vorstellung rufen*). A manifesting is only a manifesting of that which is not named, of my presentation.[15]

In other words, the name names the object *tout court* by the mediation of the content of the presentation (the object-presented-as-it-is-presented), which is expressed by the name. "The highest bell tower in Salzburg" names the bell tower of the Hohensalzburg fortress by the mediation of the content of presentation that is expressed by the name. But "the highest bell tower in Salzburg" names its object under a particular feature (namely, its height), and it leaves aside many other aspects (its architect, its date of construction, its colour, etc.). This is why Brentano says that names name their object in a more or less determinate (or indeterminate) way, similar to the more or less determinate (or indeterminate) way in which contents of presentation, the *Sinne*, present their objects. Here again, we find a similarity to Frege's account of the distinction between *Sinn* and *Bedeutung* that is not merely terminological: Brentanian and Fregean Sinne are ways of presenting or determining a *Bedeutung*, in Fregean terminology, or an object, in Brentanian terminology. They are in the same sense "modes of givenness" (*Gegebenheitsweisen*) of an object, which may or may not exist. But the interesting feature here is in the degree of determination of the contents. If the content expressed by "the highest bell tower in Salzburg" is less determinate than the content expressed by "the highest white bell tower in Salzburg," it suggests that "having a degree of determinateness" is a property only of those presentations which present their object with a minimal complexity. Such presentations are called by Brentano "abstract presentations" in contrast to the so-

---

[15]"Die Namen sind die Ausdrücke von Vorstellungen. Die Vorstellung ist aber doch nicht das, wovon man sagt, dass der Name es nenne. Der Name drückt vielmehr die Vorstellung so aus, dass er das nennt, was die Vorstellung vorstellt, und es unter ihrer Vermittlung und darum auch in derselben Weise vollstŁndig oder unvollständig bestimmt oder unbestimmt nennt, wie sie es vorstellt. So ist denn die Vorstellung zwar der Sinn des Namens; die Sache aber das durch den Namen Genannte und im eigentlichsten Sinn durch ihn Bezeichnete... Das Nennen ist kein Kundgeben des Genannten sondern nur ein in die Vorstellung rufen; ein Kundgeben ist es nur etwa von dem, was nicht genannt wird, von meiner Vorstellung." Brentano EL 72, 12578–12579 (ca. 1875–1877).

called "intuitions," in which the contents are presented as fully individualized, "at once," so to speak. According to this idea, intuitions do not have merely a degree of determinateness, but are fully determinate. Again, this might express an important distinction between content and object. What is fully determinate does not allow for a distinction between content and object: only what allows for a degree of determinateness allows for a distinction between content and object. This would also speak in favour of the strong reading of the distinction between content and object.

These few examples taken from Brentano's manuscripts on logic and psychology show that the distinction between content and object was not an idea that he merely considered at some point, but is constant from the end of the 1860s at least up to the lectures on descriptive psychology of 1890–91. In all the cases discussed, the following features of the distinction between content and object apply equally:

a) From a phenomenological perspective, there seems not to be any distinction between the content and the object of intuitive presentations, that is, of presentations which do not as such involve a linguistic description as an intrinsic part of them or any particular mental operations (such as abstraction or any other operation involving a relation between two presentation contents). From the standpoint of Brentano's phenomenology, it seems reasonable simply to accept that such intuitive presentations (which Brentano also calls sensations) do not allow for the kind of double direction of a presentation which follows from the distinction between content and object.[16] Since the absence of determinateness in intuitive presentations also speaks for their lack of double direction, it makes good sense, for phenomenological reasons at least, to reserve the distinction between content and object for mental acts of higher complexity than intuitive presentations. In this sense, the strong reading of the distinction between content and object should be preferred over the weak reading, but for phenomenological reasons. In the quotation from the lectures on descriptive psychology, he at least suggests that this distinction would hold also for unexpressed contents. In this case however, the distinction is an ontological one between two kinds of objects.

b) The distinction among the presentation, the presented-object (the content, or the object-presented-as-it-is-presented), and the object *tout court* seems to be perfectly natural in the context of the manuscripts discussed here. Why then does this distinction not play any significant role in the *Psychology from an Empirical Standpoint* (1874), which was written in a period when Brentano argued for this distinction in his manuscripts? As suggested by Fréchette

---

[16] I argue for this point in Fréchette 2016.

(2013), a closer reading of the *Psychology* from this perspective is quite revealing. First, Brentano never uses the term "content" (*Inhalt*) as a clear synonym of "object" (*Gegenstand*): he speaks of intentionality in terms of a "relation to an object" (*Beziehung auf einen Gegenstand*), of the objects of experience, of the objects of external perception, and of the external objects, but in most of the cases relevant here, "content" is not a synonym of "object." Second, when he uses the term "content," it is never in contexts where it could mean something like the object-presented-as-it-is-presented. Brentano's concern in the *Psychology* is mainly to lay the ground for a science of the mind, proposing a classification of mental acts and a description of the different kinds of mental acts: objects-presented-as-they-are-presented simply do not play a prominent role in this context, although nothing that Brentano himself wrote in the *Psychology* can be taken as supporting in any way the non-distinction view that is often attributed to him.

c) A further common point that clearly emerges from the passages I have quoted is that a phenomenological analysis of mental acts — which is the basic program of descriptive psychology (or phenomenology) according to Brentano — must take into account the way in which these acts manifests themselves. While I might discover in my experience relatively simple acts of sensing — for instance, some "Ganzfeld" experience of white while wearing special glasses that cover the whole visual field in white — most of them are more complex: take for instance my visual and auditive memory of a concert I recently attended, or my preference for Riesling over Chablis. Even for my own self-analysis, these experiences and their content must be described. For myself at least, I cannot conceive of a description which is not constitutively reliant on linguistic devices. It seems absurd to try to describe my experience of the mineral tone of a Riesling wine I drank without the help of expressions like "experience of a mineral tone" used as descriptions of the features of such an experience. In short, what these passages have in common is that they all accept that description involves the use of a linguistic device. Moreover, if description is to have any phenomenological value, it must be understood by others as well. In fact, phenomenological description itself presupposes the distinction between content and object: my phenomenological description of the taste of a good Riesling (as dry and mineral, say) has some value as a description because it is meant to capture some salient features of the object which I would like you to attend to when having a similar experience. If the content were the object, phenomenological descriptions would be a quite trivial venture in the best case, or they would be simply incommensurable with one another. Suppose you and

I are drinking the same Riesling, and we both describe the wine as mineral and dry. If the content of our experience is identical with the object, then either (a) the description "This wine is mineral and dry" is analytically true, and not a serious candidate for an example of a description; or (b) "This wine is mineral and dry" applies in fact to a private experience, is non-shareable, and cannot be subject to revision. In both cases, descriptions would turn out to be perfectly useless (in particular as phenomenological tools!) if no distinction between content and object is allowed. In taking into account the linguistic manifestations of mental acts (names and assertions, for example), Brentano seems well aware that these manifestations are the precondition for any phenomenological description.

4 Immanent intentionality and methodological phenomenalism

Let us now turn back to the strategy adopted by those readers of Brentano who are tempted to understand him as a defender of the immanent intentionality thesis, that is, the thesis that intentionality is nothing but a relation between two mental entities. I suggested earlier that such readers at some point have to characterize Brentano's conception of intentionality in one of the two following ways: (1) as representative of a more general world view, in which case they would have to attribute to Brentano some kind of phenomenalism; or (2) as a local theory of mental content which has no particular bearing on other important philosophical questions. For reasons which I have already mentioned, I doubt that strategy (2) would be realistic, given the fact that intentionality appears prominently not only in Brentano's philosophy of mind, but also in his logic, in his ethics, in his theory of knowledge, and at important places in his metaphysics. Strategy (1) has more potential, at least prima facie. One particularly interesting characterization of Brentano's conception of intentionality coming from this side is the one proposed by Peter Simons, who sees Brentano's conception of intentionality and psychology in general as driven by a "methodological phenomenalism." What is Brentano's "methodological phenomenalism"? According to Simons (1995, xvi), it is a position close to phenomenalism, with a restriction on the nature of substance, since substances exist in Brentano's metaphysics and are not reducible to phenomena, as a strong phenomenalist like Mach would argue. As a methodological phenomenalist, Brentano would argue that sounds, colours, and all other intentional objects have a purely mental existence, while also affirming that the world is ultimately made up of substances with their (ontologically dependent) accidents. If the label of methodological phenomenalism captures Brentano's world view, it seems that this world view suffers from a certain bipolarity: he would be arguing on the one hand that intentionality is a relation between mental entities, and on the other hand would be maintaining that these entities, the mental and physical phenomena, are actually accidents of a substance, which is the soul. But there seems to be no bridge between these two utterly distinct attitudes.

Another way to formulate the view would be to say that as I first-personally experience my mental life, I do not have access to any substances, but only to their phenomena. In some sense, from the point of view of one's own experience, phenomenalism might well appear to be true, but even if it were so, it would not change anything about that experience. One cannot infer from the fact that we have no access to a substance that there is no substance. This position should thus be distinguished from Berkeley's phenomenalism, and from Mach's position. However, it makes intentionality a feature that cannot reach out to anything other than what is already in consciousness. From the point of view of one's own experience, methodological phenomenalism and phenomenalism are simply indistinguishable as philosophical positions. And since intentionality is something that we have access to from the point of view of our own experience, the best option is to make the pragmatic decision of denying (without any proof) the reality of what we are directed toward. This is the gist of the idea formulated by Brentano at the beginning of the second book of the *Psychology*:

> It is not correct, therefore, to say that the assumption that there exists a physical phenomenon outside the mind which is just as real as those which we find intentionally in us, implies a contradiction. ...It is only that, when we compare one with the other we discover conflicts which clearly show that no real existence corresponds to the intentional existence in this case. And even if this applies only to the realm of our own experience, we will nevertheless make no mistake if in general we deny to physical phenomena any existence other than intentional existence (Brentano 1995, 72).

I believe that this passage illustrates my reconstruction of Brentano's alleged methodological phenomenalism:

i) In experience, we only have access to phenomena (therefore, in experience, the phenomenalist and methodological-phenomenalist standpoint are indistinguishable);

ii) given (i), we are more likely to give a correct account of the objects given in our experience when we consider them as intentionally (in)existent, and not as existent in reality.

iii) The decision is however a pragmatic one and does not involve any ontological commitment (Brentano remains substantialist in metaphysics).

For readers of Brentano who are sympathetic to the immanent intentionality view (such as Dale Jacquette), Brentano's methodological phenomenalism in this reconstruction is a welcomed confirmation of the view. According to them, intentionality is a relation between two mental entities, and has no bearing on the world. There are two serious problems with this reconstruction of Brentano's position. The first was already mentioned, and concerns the absence of a bridge between the phenomenalist reading of intentionality and the realist metaphysics defended by Brentano. If one takes the methodological-phenomenalist reconstruction seriously, psychology and metaphysics have to do with two unrelated domains and must be explained separately (the metaphysics of the mind would have no real connection with psychology). The second problem of methodological phenomenalism, however, is that it actually falls under strategy (1), which consists in the restriction of Brentano's concept of intentionality to a specific domain which has no bearing on other domains (or which involves no ontological commitment over the entities of other domains). In other words, methodological phenomenalism cannot account for the intentionality involved in communication or action. At bottom, it seems that the account cannot open up Brentano's position to anything other than the improbable conjunction of a phenomenalist psychology with a substantialist metaphysics. To be sure, the positions are not contradictory, but their conjunction suggests a world view that is far from the ideal of organicity described in section 2 above. For this reason, I believe that reconstructions of Brentano's view of intentionality that follow the path of phenomenalism are simply unable to account for the further features of intentionality discussed in section 3. What is more, they are unable to account for the kind of metaphysics advocated by Brentano in his published and unpublished works.

***

In contrast with such a strategy, I suggested that we take as a starting point Brentano's general philosophical endeavour (what I called the "world view") in order to understand how specific positions within his general endeavour are related to one another. From this general perspective, it seems simply implausible to offer a reading of intentionality based on the non-distinction between content and object.

With this fresh look at Brentano's philosophical endeavour, I then turned to the manuscripts, but also to some published works, where the distinction between content and object is obviously essential to the analysis of intentional experiences. Furthermore, the distinctions presented in the manuscripts resolve some problems in the interpretation of the theory that the immanent intentionality view leaves unaddressed. These constitute reason enough to take another look at Brentano's conception of intentionality.

The motivation behind the strategy adopted here may sound odd in some respect. In fact, what I have attempted to do, with more or less success, is to follow Brentano's own methodological approach to the history of philosophy:

> One should seek to resemble (*gleichen*) as much as possible the mind of which we want to understand the imperfectly expressed thoughts. In other words, one must prepare in advance the understanding by first encountering the philosopher philosophically, before concluding as an historian (Brentano 1911, 165). [17]

References

[1] Bergman, H. (1912), "Review of Aristoteles und seine Weltanschauung, by Franz Brentano. Quelle & Meyer, Leipzig, 1911," in Prager Tagblatt, Bd. 37, Nr. 358 (28. Dezember), 7.

[2] Brentano, F. (1874), Psychologie vom empirischen Standpunkte, Duncker und Humblot, Leipzig.

[3] Brentano, F. (1907), Untersuchungen zur Sinnespsychologie, Duncker und Humblot, Leipzig.

[4] Brentano, F. (1911), Aristoteles Lehre vom Ursprung des menschlichen Geistes. Leipzig: Veit.

[5] Brentano, F. (1959), Grundzüge der Ästhetik, Francke Verlag, Bern.

[6] Brentano, F. (1995), Psychology from an empirical Standpoint, London, Routledge.

[7] Brentano, F. EL72, Logik, Manuscript, Hougton Library, Harvard University.

[8] Brentano, F. EL81, Fragment über Logik, Manuscript, Hougton Library, Harvard University.

[9] Brentano, F. EL80, Logik, Manuscript, Hougton Library, Harvard University.

[10] Brentano, F. PS48, Vorarbeiten zur Psychognosie, Manuscript, Hougton Library, Harvard University.

---

[17] German original: "Man muss möglichst dem Geist zu gleichen suchen, dessen unvollkommen ausgesprochene Gedanken man begreifen will. Mit anderem Worte, man muss das Verständnis anbahnen, indem man, ehe man als Historiker abschließt, zunächst selbst philosophierend dem Philosophen entgegenkommt." Earlier versions of this paper have been presented at conferences in Liège, Genève, Prague, Salzburg, and São Paulo. Many thanks to Johannes Brandl, Laurent Cesalli, Arkadiusz Chrudzimski, Denis Fisette, Hynek Janoušek, Fred Kroon, Bruno Leclercq, Kevin Mulligan, Mario Porta, Robin Rollinger, Denis Seron, Hamid Taieb, and Petr Urban for their comments. This paper was written as part of the Austrian FWF project P-27215 on Franz Brentano's descriptive psychology.

[11] Brentano, F. (forthcoming), Deskriptive Psychologie und beschreibende Phänomenologie. Vorlesungen 1887/88 und 1888/89, Dordrecht, Springer.

[12] Chisholm, R. (1957), Perceiving, Ithaca, Cornell University Press.

[13] Chisholm, R. (1960), Realism and the Background of Phenomenology, New York, Glencoe.

[14] Chisholm, R. (1967), "Intentionality," in Edwards, P. (ed.), Encyclopaedia of Philosophy, London, Macmillan.

[15] Crane, T. (2006), "Brentano's concept of intentional inexistence," in M. Textor (ed.), The Austrian Contribution to Analytic Philosophy, London, Routledge, pp. 20 — 36.

[16] Fréchette, G. (2013), "Brentano's Thesis (Revisited)," in D. Fisette, G. Fréchette (eds.), Themes from Brentano, Amsterdam, Rodopi, p. 91 — 119.

[17] Fréchette, G. (2016), "Brentano on Sensory Intentionality", in Studia Philosophica, vol. 75, p. 33 — 50.

[18] Husserl, E. (1900/1), Logische Untersuchungen (in 3 volumes), Halle, Max Niemeyer.

[19] Jacquette, D. (1990), "The origins of Gegenstandstheorie: Immanent and transcendant intentional objects in Brentano, Twardowski, and Meinong," Brentano Studien, vol. 3, p. 277 — 302.

[20] Jacquette, D. (2004), "Brentano's Concept of Intentionality," in D. Jacquette (ed.), The Cambridge Companion to Brentano, Cambridge, Cambridge University Press, p. 98 — 130.

[21] Kriegel, U. (forthcoming), Brentano's philosophical system, Oxford, Oxford University Press.

[22] Marek, J. (forthcoming), "Brentano und Meinong. Versuch einer Gegenüberstellung".

[23] Meinong, A., Höfler, A. (1890), Logik, Vienna, Tempsky.

[24] Meinong, A. (1899), "Über Gegenstände höherer Ordnung und deren Verhältnis zur inneren Wahrnehmung." in Zeitschrift für Psychologie und Physiologie der Sinnesorgane, vol. 21, p. 182 — 272.

[25] Simons, P. (1995), "Introduction to the Second Edition", in Brentano, F. (1995), p. xii — xx.

[26] Smith, B. (1995), Austrian Philosophy. The Legacy of Franz Brentano, Chicago, Open Court.

[27] Twardowski, K. (1894/1977), Zur Lehre vom Inhalt und Gegenstand der Vorstellungen. Eine psychologische Untersuchung, Vienna, Hölder. English translation: On the Content and Object of Presentations. A Psychological Investigation, translated by R. Grossmann, The Hague, Martinus Nijhoff.

Nuclear and Extra-nuclear Properties

Nicholas Griffin
McMaster University, Canada.
ngriffin@mcmaster.ca

The last quarter of the last century saw a remarkable revival in Meinong's reputation as a philosopher. His theory of objects, which for the previous half century or more had been held up to ridicule as a paradigm of how not to do philosophy, became the subject of serious philosophical research and the inspiration for a variety of neo-Meinongian semantic theories. Dale Jacquette was one of the most prolific contributors to this development in a series of books and articles going back to the 1980's and originating in his 1983 doctoral dissertation, "The Object Theory Logic of Intention", written at Brown with Roderick Chisholm as supervisor — though Chisholm's influence on Jacquette's published work on Meinong does not appear to be great.[1] One of the things that distinguished Jacquette from other neo-Meinongians of the same period was the detailed attention he paid to Meinong's own views. The neo-Meinongians all deviated from Meinong's own position in various ways and Jacquette was no exception, but while the others tended to refer back to Meinong only for odd points of detail or as an initial source of inspiration, Jacquette, in addition to developing his own neo-Meinongian semantics, also contributed historical and exegetic articles, work which other neo-Meinongians have tended to avoid. Exegetic work on Meinong is not for the faint-hearted, as Jacquette himself acknowledged in an autobiographical piece on his life as a Meinongian logician, where he noted that his German was good enough for Schopenhauer and Wittgenstein, but was 'overwhelmed ... by Meinong's ... tortured philosophical prose' ([2000], p. 371). Even translation doesn't help,[2] for Meinong came to his mature theory of objects bit by bit via a tortuous evolution which involved an exhausting examination of all

---

[1] Not long before his death he collected many of his writings on Meinong together in a single volume (Jacquette [2015a]), which is exceptionally useful for the commentator. My citations are to this volume wherever possible, though it should be noted that Jacquette revised some of his papers when reprinting them.

[2] Years ago, I paid a bi-lingual graduate student to work on translations of some key passages for me. The work was exhausting for both of us, and painfully slow. His final judgement was, 'I know what it says, but I have no idea what it means' — and neither did I. It became clear that, at least as far as I was concerned, the practicable route to a better understanding of the theory of objects did not lie through a close study of Meinong's texts.

manner of dead ends and false starts, his final position seeming to emerge (not too clearly) as what was still left when all other options had been laboriously abandoned.

The late-twentieth-century upsurge of interest in Meinong's theory of objects did not result from a renewed interest in Meinong's own work, which lay, like the Sleeping Beauty, protected from all advances behind the impenetrable thicket of Meinong's prose. It resulted rather from the insuperable difficulties encountered by the hitherto prevailing semantic tradition, namely the referentialist tradition initiated by Russell and centred on his theory of definite descriptions, when it attempted to deal with intentional discourse.[3] It proved incredibly difficult — I would maintain, impossible — to find an adequate semantics for intentionality within the referentialist framework, which confined all genuine reference to existent objects and yet, in dealing with intentionality, had to provide a treatment of language which was putatively about the imaginary, the fabulous, the absurd, and the paradoxical. None of the accounts referentialists offered was able to capture the content of such discourse adequately. Meinong's theory of objects offered to do just that, and in a particularly simple and straight-forward way. Meinong held that all intentional discourse about objects which did not exist, discourse which referentialists had to find some way of analyzing away, did exactly what it appeared to do: it referred to genuine objects, but ones which did not exist, and objects which, moreover, had (broadly speaking) the properties that were ascribed to them. He implemented this approach through the two fundamental principles of his theory of objects: the principle of unlimited freedom of assumption:

(UFA) However we characterize an object, there is an object thus characterized;

and what Routley ([1980], p. 46) called the characterization postulate:

(UCP) An object has all those properties which characterize it.[4]

The theory which results from these two principles — unqualified, straight out of the package as it were — is the naive theory of objects. It is beautiful, straight-forward, and simple, but it is wrong, and it was not Meinong's final position: it was more like a first approximation which had to be revised and qualified before it could be defended.

---

[3] I am not claiming that referentialism's problems were confined to intentional discourse, merely that that was where they were most egregious. Jacquette briefly surveys Meinong's changing fortunes in Jacquette [2015e].

[4] To my knowledge, Meinong does not state the characterization postulate as a principle; nonetheless, he does appeal to it implicitly in his treatment of examples (e.g., Meinong [1904], p. 82). Unlimited freedom of assumption, on the other hand, is explicitly stated (Meinong [1915], p. 282).

It is pretty clear that the naive theory of objects won't do. It is subject to two devastating objections raised by Russell almost as soon as Meinong put it forward. Russell argued, first, that, Meinong's theory violated the law of non-contradiction, since according to it the round square is round and also not round; and, second, that the theory entails patent falsehoods such as that the existent golden mountain exists.[5] Both arguments stem directly from the two founding principles of naive object theory. Since we can characterize an object as a round square or as an existent golden mountain, it follows by (UFA) that there are such objects, and by (UCP) that the former is round and square and that the latter exists. By an unstated but not unreasonable step, Russell concludes that the former is both round and not round and thus violates the law of non-contradiction. I shall return to this argument later, but for the moment it is the other argument that I want to look at. The problem that the second argument poses for the theory of objects is that, given both (UFA) and (UCP) and any set of properties Γ, we can characterize an object which has all and only the properties in Γ. If, then, we add existence to Γ, we can characterize another object which has all the previous properties and which also exists. This is intolerable: obviously objects can't be called into existence simply by adding the word 'existent' to their description. The long eclipse of Meinong's theory of objects — insofar as it had a logical rather than a merely historical basis — was largely due to these two arguments.

Meinong quickly responded to Russell's arguments. To the first, he replied that the law of non-contradiction does not apply to impossible objects: indeed, they were impossible precisely because they had inconsistent properties (Meinong [1907], pp. 14-15). Against the second argument, he acknowledged that one could add what he called an 'existential determination' to the set of properties which characterize an object and thus one 'can just as much speak of an "existent golden mountain" as of a "high golden mountain", and then assert the predicate "existent" of the former just as readily as "high" of the latter'. Nonetheless, ' "to be existent" in the sense of an existential determination and "to exist" in the ordinary sense of "being there" are not at all the same thing' (ibid, p. 18; also [1910], p. 148). So, according to Meinong, the existent golden mountain is existent, but nonetheless does not exist, is not there. Not surprisingly, Russell, in another review, said that he could not see the difference between being existent and existing (Russell [1907], p. 692; also [1910], p. 157). It does, indeed, seem to be a distinction in need of a difference.

Eight years later, Meinong tried again, this time making a distinction between two different types of property: *konstitutorisch* and *ausserkonstitutorisch* (Meinong [1915], pp. 176-7). Findlay ([1933], p. 176) introduced 'nuclear' and 'extra-nuclear'

---

[5]Russell [1905], p. 418; [1905a], pp. 598-9; [1907] pp. 92-3.

as translations for these terms and his usage has stuck.[6] Meinong now claimed that only nuclear properties can be used to characterize an object, the extra-nuclear ones cannot. The idea is that an object is characterized by 'ordinary', nuclear properties, like being golden and being a mountain and by these alone. Existence, on the other hand, is not an ordinary property and cannot be used to characterize an object. Existence is thus among an object's extra-nuclear properties. A non-existent object has the nuclear properties which characterize it, but not the extra-nuclear ones which don't. Thus the existent golden mountain is golden and is a mountain, but does not exist. Strictly, there are two options open to the Meinongian here, depending upon which of the two fundamental principles, (UFA) and (UCP), they restrict to nuclear properties only. One could maintain that only nuclear properties can characterize an object, but the object has all the properties which characterize it, thus restricting (UFA) but not (UCP). Alternatively, one could hold that all properties, even extra-nuclear ones, may characterize an object, but the object only has the nuclear properties which characterize it, thus restricting (UCP) but not (UFA). Meinongians have rarely been explicit about which of these options they prefer. On the second option, there is the problem of explaining the relation that an object has to the extra-nuclear properties which characterize it: can an object really be characterized by a property which it doesn't have? But the first is not free from problems either, as we shall see. Either way, Meinong's distinction between nuclear and extra-nuclear properties has commended itself, in some form or another, to almost all subsequent Meinongians.

The idea that existence is not an ordinary property has a distinguished philosophical pedigree, at least since Kant's refutation of the ontological argument (Kant [1781], A598-600 = B626-28). And Kant is often cited, if not always with full approval, by contemporary Meinongians as evidence of the reasonableness of the nuclear/extra-nuclear distinction. The traditional reading of Kant is that he simply denied that existence was a property, thus putting him in line with modern, classical

---

[6]Since it is familiar I shall continue to use it, but it is not ideal since 'nuclear' suggests a set of core or essential properties on which the other properties are anchored. 'Constitutive' would be better since an object's nuclear properties, as we shall see, are generally taken to be those which determine its identity, they constitute it as the item it is. Findlay, however, needed to distinguish '*konstitutorisch*' from '*konstitutiv*', to mark another Meinongian distinction, and (alas) chose 'constitutive' for the latter (Findlay [1933], p. 176n). Jacquette in [2001] (and elsewhere) used both 'nuclear' / 'extranuclear' and 'constitutive' / 'extraconstitutive', and in reprinting [2001] in Jacquette [2015a] even changed the title to include both. The distinction was originally introduced by Ernst Mally [1908] as one between formal ('*formalen*') and extra-formal ('*ausserformalen*') properties. Mally's terminology seems less odd in the context of the mathematical example he was discussing. Meinong ([1915], p. 176) noted its inappropriateness outside that context. For more on Mally's innovation including a translation of the key texts, see Jacquette [2008].

quantification theory which treats existence by means of existential quantification. But this interpretation of Kant is by no means obvious and Jacquette ([2001], p. 86-88), for one, rejected it.[7] Jacquette held, rather, that for Kant 'exists' was a special kind of predicate, not a 'determining predicate', that is, not in Kant's words 'a predicate which is added to the concept of the subject and enlarges it' (A598 = B626).[8] As Jacquette goes on to explain, the concept of a 100 existent gold Thalers is exactly the same as the concept of 100 gold Thalers or 100 non-existent gold Thalers — the addition of 'existent' or 'non-existent' to its specification makes no difference to the concept nor to its object, which is the same in all three cases. On Jacquette's account, the properties of being golden, being 100 in number or being a Thaler (nuclear properties for the Meinongian) are, for Kant, determining properties, which alone, in Kant's terms, constitute the concept and individuate its object. In this respect, Kant's determining properties and the Meinongians' nuclear properties play the same role. But it's the other side of the distinction that makes the Meinongians' appeal to a Kantian distinction between determining and non-determining properties problematic as a template for their own distinction between nuclear and extra-nuclear properties. So far as I know, Kant says essentially nothing about non-determining predicates. In his system there seems little for them to do. Indeed, on Kant's account, existence seems to be not so much a property as, what Kemp Smith ([1923], p. 530) calls, a 'purely formal factor' which, as Kant puts it, 'posit[s] the subject in itself with all its predicates ... as being an object' (A599 = B627). This will not do for the Meinongian, for whom every subject with all its predicates is posited as an object: there is no need for a 'formal factor' to establish its objecthood. For the Meinongian, 'exists', just like any other predicate, picks out a relatively small subset of the immense domain of objects. If objects are to be posited in Meinongian systems, it is by means of ontologically neutral quantifiers which range over the entire domain (cf. e.g., Jacquette [1996]; [2009], p. 140). So the reasons to which Kant can appeal in order to justify claiming either that 'exists' is not a predicate or that it is not a determining predicate are simply not available to Meinongians to support their claim that 'exists' is not a nuclear predicate.

But even if they were, they would hardly be sufficient to fund the entire distinction between nuclear and extra-nuclear properties. For Kant's case is specific

---

[7] Others include Routley, [1980], pp. 180-1; Campbell [1974]; Allison [1983], p. 342n23; Wood [1992], p. 400.

[8] What Kant actually says is that 'being' (*Sein*) is 'obviously not a real [*reales*] predicate; that is, it is not a concept of something which could be added to the concept of a thing [i.e., a determining predicate] ... If ... [we] say "God is", or "There is a God", we attach no new predicate to the concept of God, but only posit the subject in itself with all its predicates, and indeed posit it as being an *object* that stands in relation to my *concept*.' (A598-9 = B626-7). The case, like much in Kant, is not as clear cut as one might wish.

to existence itself, and existence is by no means the only extra-nuclear property. It is not even the extra-nuclear property that Meinong was discussing when he introduced the distinction. The properties he had in mind were simplicity and determinateness; before him, the example Mally used was completeness. Kant's case for the exceptionalism of 'exists' cannot be made against these three. In the case of simplicity, determinateness and completeness — at least as Mally and Meinong understood them — one might make a case for their exceptionalism on the grounds that they are higher-order properties, which supervene on first-order properties.[9] One might then perhaps venture that first-order properties were nuclear and higher-order properties extra-nuclear. And one can make a case for including existence as an extra-nuclear property by treating it as higher-order,[10] though it's by no means obvious that it is higher-order or that Meinongians should treat it as such. Be that as it may, it is hard to think that all higher-order properties are extra-nuclear. The parallelogram-*in-abstracto*, to use Mally's example (Jacquette, [2008], p. 397), has the higher-order property of having all the defining properties of a parallelogram. But there is no reason to think that this is an extra-nuclear property, and no problem arises from using it to characterize an object: it is the natural way to do so for someone who has forgotten what the defining properties of a parallelogram are. Whether extra-nuclear properties constitute some proper subset of higher-order properties is, I think, up for debate and, with a certain amount of gerrymandering (e.g., with 'exists'), may be admissible. The gerrymandering itself is not necessarily a problem: the distinction between nuclear and extra-nuclear properties is an important one for Meinongians and they may legitimately tailor their accounts of tricky properties to preserve a plausible way of making the distinction against apparent counter-examples. But, as things stand, all this is conjectural since, so far as I am aware, no satisfactory demarcation line among higher-order properties as been proposed.

Meinongians have always drawn the distinction between nuclear and extra-nuclear properties inductively, by specifying clear examples of extra-nuclear properties and then generalizing. The initial examples are identified by their ability to generate obvious falsehoods when the two fundamental principles of object theory, (UFA) and (UCP), are applied. We have seen how this works in the case of existence, but consider the example Meinong uses, determinateness. He considers characterizing an object as 'a square brown thing' in a case where the object intended is a completely determinate object, that is an object with a full range of properties which distinguish it from everything else (Meinong [1915], p. 189). By the characterization postulate, the object characterized has just the properties of being

---

[9]Meinong makes essentially this point ([1915], p. 189).

[10]Such a move, which goes back to Frege, is not inherently implausible and has been taken up by some Meinongians, e.g. Routley ([1980], pp. 244-8), though not for the reasons canvassed here.

brown and being square and thus is distinct from the determinate object intended. Meinong considers whether we might get around this by characterizing the object as 'a determinate square brown thing'. But this, he notes, will not do at all. By the characterization postulate, the object now characterized has the properties of being brown, being square and being determinate, but this is still not the object intended, for it is still incomplete with respect to all the other properties which make the intended object determinate, like being large or hard or heavy. As Findlay ([1933], pp. 175-6) puts it 'we have left quite undetermined in what ways the square brown thing is determinate', and this because determinateness is 'not really a property of the same simple type as squareness, brownness or heaviness. The determinateness or indeterminateness of an object is not a part of its nature in the same immediate way that the other properties are.' Meinong ultimately resolves this issue by distinguishing between an auxiliary object (*Hilfsgegenstand*) and an ultimate object (*Zielgegenstand*): it is by means of the indeterminate auxiliary object that we refer to the determinate ultimate object (Meinong, [1915], p. 196). The account does not look terribly plausible — though perhaps no less so than Frege's account of how a sense presents a reference — but the details do not concern us here. The point is that characterizing an object as determinate does not make it so, whatever the characterization postulate says.

The same goes for all the other extra-nuclear predicates. There are many of them. Elsewhere, I've called them 'status predicates', since they all seem to indicate an object's standing rather than its nature (Griffin [2008]), but this merely gives them another label, without identifying the feature(s) which make it legitimate to group them together. They include ontological predicates (in addition to 'exists', 'is real', 'is actual', 'is fictional', 'is mythological'), modal predicates ('is possible', 'is necessary', 'is contingent'), logical predicates ('is consistent', 'is contradictory', 'is complete', 'is valid'), and many more, together with their opposites, for one clear general principle in the area seems to be that, for any predicate 'F', if 'F' is extra-nuclear, so too is 'not-F'. There is not much disagreement among Meinongians as to which predicates are extra-nuclear, at least for a large range of central cases, and that has led to the feeling that some intuitive, philosophically significant criterion will mark them out. But the criterion itself has proven elusive.

Jacquette, to his credit, took this bull by the horns and attempted to give a formal definition of extra-nuclear properties (Jacquette [2001]). First, however, he looks at three previous attempts and finds them wanting. He starts with Findlay. Findlay does not explicitly state a criterion, but his treatment of the simplicity example suggests one — or rather two (Findlay [1933], p. 176). He considers a simple object, a specific shade of red, which he understands as having just the one (nuclear) property of being that particular shade. If we characterize this object as simple, i.e. if we treat its simplicity 'as part of the nature of the shade', we get a contradiction, for then the object has two properties, simplicity and redness, and thus is complex. Findlay also has a second argument: 'if simplicity be an element in the shade of red, all objects that are characterized by the shade will be also characterized by simplicity, which is absurd' (ibid.). The implicit criterion seems to be that, if characterizing an object as F results in contradiction or absurdity, 'F' is an extra-nuclear predicate.

There's lots that's wrong with this. To take the argument to absurdity first: it is, as Jacquette points out, 'an evident *non sequitur*': red is a colour, and blood is red, but blood is not a colour, so the general principle on which the argument rests is false ([2001], p. 91). Moreover, Jacquette's counter-example to the principle suggests a problem for the argument to contradiction as well. For the shade not only has the property of being red, it also has the property of being coloured (and, for that matter, the property of being the same colour as blood), and so it was never simple in the first place. If the argument to contradiction is to work, it is essential that 'being simple' is understood as 'having just one property'. But the shade is not simple in that sense: merely in virtue of having the property of being red, it has many additional properties; and when negative properties are taken into account, it has a host of them (it is not-green, for example). 'Perhaps', Jacquette says, 'there are no objects whose nature is simple in the required sense' (ibid.). But that is not right either. For, if freedom of assumption is unlimited, we can characterize an object as having just one property F (indeed, we are doing so now); by (UFA) there will be such an object and by (UCP) it will have just the one property. It will, however, be a very peculiar object, for in addition to F, which by (UCP) is the only property it has, it will also have the properties of being characterized by F and of having F as its only property. It seems the object is a paradoxical one, but that, in itself, is no reason for object theory to shun it, any more than it should shun the round square — there are more ways of being inconsistent than simply being characterized by incompatible properties. At any rate, this object is something very different from Findlay's specific shade of red which was intended to be a relatively ordinary phenomenal object. The specific shade of red was never simple in the required sense of having only one property.

Jacquette's subsequent objections to the argument to contradiction are more telling. He points out that, if the purpose of the nuclear/extra-nuclear distinction is merely to avoid contradiction when an object is characterized as red and simple, this can be achieved as readily by categorizing redness as extra-nuclear and simplicity as nuclear as the other way around. The argument to contradiction cannot, on its own, show that simplicity is the offending property, some further ground against treating simplicity as nuclear is needed (Jacquette [2001], pp. 91-2).[11] Moreover, and more importantly, it is not enough to point out that these problematic simple objects lead to contradictions, for Meinong clearly intended that objects could be characterized in putatively inconsistent ways. The round square, after all, is both round and square, and Meinong (as we've seen) did not, at least initially, flinch from accepting that it involved a contradiction. If the argument to contradiction is to show that simplicity is not a characterizing property, Findlay would need to show that the contradiction which resulted from treating it as one was in some way more damaging than that produced by the round square (Jacquette, [2001], p. 92). Whether Meinong was, ultimately, quite so sanguine about the round square as he appeared to be in his initial response to Russell will have to wait until we turn to his later response to Russell's first argument.

Routley ([1980], pp. 264-8) offers by far the most extended account of the nuclear/extra-nuclear distinction. He uses the terms 'characterizing' and 'non-characterizing' and does not assume that his distinction is co-extensive with the original Mally-Meinong one. He lists a whole series of distinctions between different types of predicates which have historically been made by philosophers and logicians and which bear some similarity to the characterizing / non-characterizing distinction. The Meinong-Mally distinction is just one of these — albeit one that is, Routley says, 'especially germane' (ibid., p. 265).[12] These earlier distinctions do give some credibility to Routley's claim that the nuclear/extra-nuclear distinction is 'intuitive and rather natural' (ibid. p. 264). We can hardly say that they draw the same demarcation line (none of them is precise enough), but we can say that they all smudge in the same area. But, as Routley acknowledges, our intuitive feeling that there is some important distinction to be made here does not obviate the need to 'elaborate the distinction and to try to make it good' if the theory of objects is to be developed, applied, and assessed (ibid.).

At this point, it would be nice to have necessary and sufficient conditions, but

---

[11]Parsons ([1980], p. 23n) notices the problem and credits Dorothy Grover for the following argument in connection with existence. Suppose existence were nuclear, then freedom of assumption will give us an object which has that property and no other. Such an object would be incomplete and (on standard Meinongian theories) no incomplete object exists, contradicting the original assumption. But, as is often the case in connection with extra-nuclear properties, it is not clear how it can be generalized. We can, indeed, extend it to simplicity: If simplicity is nuclear then we can characterize an object as having that property and no other. But such an object is also incomplete, thus contradicting the original assumption. But obviously we can't use this same argument to show that incompleteness is extra-nuclear.

[12]When it comes to Parsons' subsequent systematization of the Mally-Meinong distinction, Routley notes divergences from the distinction he wants, especially as regards relational predicates (ibid., p. 265n).

these are not forthcoming. Instead, Routley offers a 'rough nonexhaustive typology of predicates' (p. 264) which 'can serve as base cases in a quasi-inductive elaboration of the distinction' (p. 265). He starts with 'descriptive predicates' as paradigm cases of characterizing predicates and 'ontic predicates' as paradigms of non-characterizing ones — an initial move which follows Kant's distinction between determining and non-determining predicates. Almost everything which follows hangs on his account of descriptive predicates, namely, 'predicates that would unobjectionably be used in describing or classifying a thing, or in older terms giving its essence or specifying its nature' (p. 265). He enumerates (nonexhaustively) various syntactical kinds of such predicates: intransitive descriptive verbs and transitive verbs (especially auxiliaries like 'to be') concatenated with descriptive adjectives, indefinite descriptions where the noun is a descriptive sortal, or a descriptive mass term.[13] These, then, constitute roughly the base class of characterizing predicates. By way of inductive extension: the predicate negation of a characterizing predicate is characterizing,[14] and so, too, are at least some compound predicates where all component predicates are characterizing (e.g., if F is characterizing and G is characterizing then so too is their conjunction).

If the class of characterizing predicates consists essentially of descriptive predicates, the class of non-characterizing predicates is much more heterogeneous. It includes, of course, ontic predicates, together with evaluative predicates (which supervene on descriptive ones), theoretical predicates (like 'is simple', 'is complete'), logical predicates (notably including identity determinates, e.g., 'is identical to a'[15]), and intentional predicates (e.g., 'is thought about', 'is searched for'[16]). As Jacquette says: 'This is certainly a more complete enumeration than anything we find in other sources. What is lacking, nevertheless, is a principle by which predicates in any of these subtypes can be understood as belonging to the one category rather than the other.' ([2001], p. 95). Moreover, there are other problems, especially as regards Routley's reliance on the undefined basic notion of a descriptive predicate to demarcate characterizing properties.

As Jacquette notes (ibid. p. 94), Routley says nothing about why, on his classification, ontic properties should not be treated as descriptive, and thus characterizing. Obviously, if we do so treat them, we will run into Russell's objection to object the-

---

[13] He does not mention mass terms, but would surely want to include them.

[14] There will be more to say about predicate negation shortly.

[15] Presumably Routley would want to restrict this to cases where a exists. There is nothing wrong with characterizing Holmes's best friend as identical to Watson..

[16] Jacquette ([2015b]) calls these 'converse intentional predicates', following Chisholm [1982]. They are especially important for the theory of objects, which has good prospects of offering a simpler and more plausible account of them than anything on offer from the referentialist tradition.

ory. But our purpose is to find a principled way of blocking that objection, not to classify any predicate which opens object theory to it as non-descriptive. In many ways, 'exists' would seem to be straightforwardly descriptive, especially in the context of Meinong's theory of objects, where it plays exactly the sort of classificatory role that descriptive predicates, on Routley's account, are supposed to.[17] The 'older terms' to which Routley also alludes, namely, specifying the nature or essence of an object, might be more helpful in showing that 'exists' is not classificatory, but for the fact that, in order to accomplish this task, they have to be underwritten by an exclusive metaphysical distinction between essence and existence which is not only controversial but not markedly clearer than the distinction is it called upon to underwrite.[18] Moreover, as Jacquette points out, in a historical novel like *War and Peace*, some of the characters (e.g. Napoleon) are plausibly described as existing and others (e.g., Bezuhov) as not. The issue here, to anticipate a later theme, is that Napoleon exists and Bezuhov does not, because the former exists in the actual world and the latter does not; in the world of the novel, both exist. In some fictions, however, the characters have a more complex relationship to existence than this. In the James Stewart movie, *Harvey*, for example, there is a real question as to whether the rabbit Harvey exists or not. This is not a question as to whether Harvey exists in the actual world (he obviously does not), but whether he exists in the world of the movie. It is hard to see why this question is not properly said to be the question of whether Harvey is correctly described as existing in the world portrayed by the film. It does not involve a puzzle about the relation of the rabbit to the actual world, but is simply one of how the rabbit is to be described in the film.

Other ontic predicates are also problematic. Routley takes as 'representative' of ontic predicates 'those predicates such that they or their negations imply existence or its negation' ([1980], p. 266). Thus he includes 'is created' and 'dies'. But that the monster in *Frankenstein* is created would seem to be a simple description of it, implying that the creature exists in the world of the novel, but not outside. Similarly, whether or not Holmes dies at the Reichenbach Falls is surely a matter of how Conan Doyle describes him. Some of Routley's inclusions are not problematic on the existence-implying criterion: 'is possible', e.g., is included since its negation implies non-existence. But he has to resort to fiat to include 'is contingent', since neither it nor its negation imply existence. Worse still, on classical quantification theory any predicate will be ontic on Routley's criterion, since on the standard semantics non-existents have no properties (Jacquette [2001], p. 94). The root problem is

---

[17]Indeed, as Routley himself argues, '*a* exists' tells us something about *a* ([1980]. p. 121).

[18]It is worth noting that Meinong held that metaphysics was only a proper part of the much more general theory of objects (Meinong, [1904], pp. 106ff). A basic object-theoretic distinction should thus not depend upon specific metaphysical doctrines.

that Routley's classification of his examples as nuclear or extra-nuclear depends almost entirely upon his classification of predicates as descriptive or not, but as Jacquette notes (ibid.), 'descriptive' is essentially synonymous with 'characterizing' in Routley's usage.

The widespread reliance on examples to make the nuclear/extra-nuclear distinction is exploited by Terence Parsons (who gives several familiar examples, [1980], p. 23) to yield, not a criterion, but a decision procedure for dividing predicates into the two categories (albeit 'a very imperfect one', and one he probably does not expect us to take too seriously) :

> [I]f everyone agrees that the predicate stands for an ordinary property of individuals, then it is a nuclear predicate and stands for a nuclear property. On the other hand, if everyone agrees that it doesn't stand for an ordinary property of individuals (for whatever reason), or if there is a history of controversy about whether it stands for a property of individuals, then it is an extranuclear predicate, and does not stand for a nuclear property. (Parsons [1980], p. 24)

Jacquette ([2001], p. 92) complains that 'ordinary' is here being used in an extraordinary sense, for existence has a good claim to be thought of as a perfectly ordinary property of individuals, whereas properties like being a unicorn, or a round square, or a prime number with three factors are, on the face of it, not ordinary at all. In reply Parsons can point to the long history of controversy about 'exists' from Kant to Frege and Russell. But he would be wrong to claim that no such controversy attended the other examples, especially the inconsistent predicates, which have often been thought not to express genuine properties.[19] Moreover, Parsons' decision procedure concedes a predicate to be extra-nuclear as soon as a challenge to its being nuclear is raised, no matter what the outcome. There is surely no good reason to suppose that any challenge is always correct, that no challenge can be beaten back. Suppose the challengers themselves admit their error (Jacquette, [2001], p. 93). On Parsons' decision procedure the predicate remains extra-nuclear.

To be fair Parsons' account does deliver four sufficient conditions for a predicate's being extra-nuclear (I combine them into one for brevity, marking the options with parentheses):

---

[19]Cf. e.g., Locke ([1690], III.iii.19) who holds that general terms stand for complex abstract ideas provided they contain no inconsistency. 'Unicorn' he specifically admits, but 'round square' and 'prime number with three factors' would surely fail his 'clear and distinct ideas' test (ibid., III.x.2). Indeed, a Lockean might challenge even redness as an 'ordinary property of individuals' since it is a secondary quality.

> F is extra-nuclear if there is a set X of nuclear properties, not containing F, such that [it is possible that] every object which has every member of X has [lacks] F.

Taking the non-modal case first, the idea is that, since every set of nuclear properties corresponds to an object which has just those properties, if X is any set of nuclear properties and $F \notin X$ then, if F were a nuclear property, there would be an object which had just the properties in X and lacked F and there would also be another object which had all the properties in X and had F as well and had no other properties. In the modal case, let X contain the single property of being a unicorn and consider a possible world which contains no myths, stories or thoughts of unicorns and in which unicorns are unreal. In that world, every object which is a unicorn will lack the properties of being mythological, fictional, and thought about and will have the property of being unreal. By Parsons' condition, it follows that these four properties are extra-nuclear (Parsons, [1980], pp. 24-5).

Of course, Parsons' conditions cannot be taken as a definition of 'extra-nuclear', for they presuppose we can already specify X as being a set of nuclear properties. But that in itself would not matter very much, for we could entirely plausibly select some (perhaps small) set of properties as paradigms of nuclear properties and then use Parsons' conditions to extend the set, property by property, in a genuine induction. The extra-nuclear properties are those which are left out. This, however, will not do because it supposes that, in the absence of a robust general distinction between nuclear and extra-nuclear properties, we can still tell which sets of properties are correlated uniquely to an object. The supposition is that if there is a unique object with exactly the properties in $X \cup \{F\}$ then F is nuclear. But how are we to tell if there is such an object? These are, for the most part, non-existent objects, we can't go out and look for them. Whether there is such an object depends upon whether the theory of objects delivers it, and this in turn depends upon whether F is nuclear. Since the properties in X are assumed to be nuclear we know that there is a unique object with exactly the properties in X. Whether the properties in $X \cup \{F\}$ pick out a different one depends, from an object-theoretic point of view, entirely on whether F is nuclear: they do if it is, they don't if it isn't. For example, we simply have no way of telling whether the existing golden mountain is a different object from the golden mountain independently of whether existence is a nuclear predicate or not. On Parson's theory it is, if existence is nuclear; otherwise, it is not. Parsons concedes that his conditions 'will not work if applied from a perspective of complete skepticism' (p. 25n), but the situation is worse than that: it seems that, if the conditions are to work, they have to presuppose the entire distinction they are expected to elaborate.

Probably the best rule of thumb we can take away from Parsons' account is not the one he actually gives, but rather: If in doubt, treat a predicate as extra-nuclear. This has the advantage of safety: wherever there is a danger that a predicate will lead to untoward results, like the ones Russell pointed out, it will be classified as extra-nuclear and the danger will be averted. But it comes with a cost. One of the signature achievements of Meinong's theory of objects was his principle of unlimited freedom of assumption, it was this that made his theory such a notable improvement on referentialist semantic theories, especially in the treatment of intentional discourse. But since extra-nuclear predicates cannot be used to characterize an object, every extra-nuclear predicate constitutes a restriction on freedom of assumption. Anyone who wants to stay true to the original inspiration of Meinong's theory will want to limit the number of extra-nuclear predicates as much as possible. Parsons attempts to mitigate the problem by maintaining that every non-characterizing extra-nuclear predicate has a 'watered-down' nuclear companion which is characterizing. Thus extra-nuclear predicates are not entirely lost for free assumption, since they can be replaced by nuclear surrogates which are (so it is implied) just as good (Parsons, [1980], p. 68). This idea and the watering-down terminology come from Meinong.

In *Über Möglichkeit und Wahrscheinlichkeit*, having introduced the distinction between nuclear and extra-nuclear properties and restricted characterization to nuclear properties only, Meinong tried to claw back what had been lost by postulating, as Parsons does, that each extra-nuclear property has a watered-down (*depotenzierte*) nuclear variant (Meinong [1915], p. 266). He backs this proposal up with an elaborate and extremely obscure account of what constitutes the difference between a normal extra-nuclear property and its watered-down companion. His view is that the former has what he calls the 'modal moment' (*das Modalmoment*), a 'full-strength factuality' which the latter lacks. Meinong says that we can use an extra-nuclear property to characterize an item, but only, he says, with a grain of salt which 'makes itself felt due to the fact that it is beyond my power to genuinely reach an actually existing object by deliberating upon its nature (*Beschaffenheit*). My authority inevitably lags behind at exactly the point of the modal moment' ([1915], p. 282).[20] The account seems to be that extra-nuclear predicates have a content, which can be used to characterize, and then something like a force, the modal moment, an attributer of status, which cannot. In using extra-nuclear predicates to characterize we can correctly attribute the content, but not the status. In characterizing an object I pick out the object which has those properties with which

---

[20]Meinong's account of the distinction ([1915], Ch. 37) is notably obscure. Findlay ([1933], Ch. 4) is more helpful. Much useful clarification is provided by Routley's (surprisingly sympathetic) critique ([1980], pp. 860-3).

I characterize it, but if I characterize it as existing (which I may) I can't, just by so characterizing it, pick out an object which actually exists, for there may be no such thing, even though I pick out an object with the character of existing. This doctrine was to some degree anticipated in Meinong's 1907 distinction between 'exists' and 'existent' which so mystified Russell. Without the distinction between nuclear and extra-nuclear properties, Meinong had no principled way of expressing the matter and it was left to look as if his distinction was a merely grammatical one between the verb 'exists' and its adjective 'existent'. In 1915, armed with the distinction between nuclear and extra-nuclear properties, Meinong revisited his reply to Russell ([1915], pp. 278-82). The early distinction is to be understood as a distinction between the 'full-strength' predicate, 'exists', and its watered-down companion, 'is existent'. The latter, which Meinong in [1907] called an 'existential determination', is a nuclear property which can be used to characterize an object; the former is not. It follows that the existent golden mountain has the watered-down property of being existent, but does not have its full-strength counterpart, existence. An extra-nuclear predicate stripped of its modal moment becomes a watered-down nuclear predicate.

It is fair to say that no one likes Meinong's doctrine of the modal moment. Most contemporary Meinongians simply ignore it. Routley ([1980], pp. 496, 863) points out that there is no need for it, since the distinction between nuclear and extra-nuclear predicates, if properly handled, will solve Russell's problems without it. Findlay, after a careful and sympathetic account, goes on to prefer an alternative theory by Mally (Findlay, [1933], pp. 110-12, 182-4).[21] Even Jacquette, perhaps the most Meinongian of all the neo-Meinongians, devoted a paper to attacking it (Jacquette [1985]; see also Jacquette [1996], p. 89; [2001], pp. 98-102). And Parsons, even though he introduces a watering-down function on predicates into his formalism, admits that he doesn't understand Meinong's modal moment (Parsons [1980], p. 44). Yet, without the modal moment, the role of the watering-down function has no explanation and neither, more seriously, does the mitigation that the availability of watered-down versions of extra-nuclear properties is supposed to provide for the damage done to the basic object-theoretic principles (UFA) and (UCP) by the exclusion of extra-nuclear properties from their scope. For a mitigation to be real there has to be some reasonable explanation of how it makes up (at least somewhat) for what has been lost. Meinong's modal moment, for all its obscurity, was at least an attempt to provide such an explanation and I shall attempt to make more sense of it later.

Jacquette ([2001], pp. 95-8) offers his own account of the nuclear/extra-nuclear

---

[21] Cf. Mally [1912]. Jacquette [1989] calls this 'Mally's heresy'. It has been revived and elaborated by Rapaport [1978] and Zalta [1983]. More details are given below.

distinction — indeed, he is the only author I know of who goes so far as to offer necessary and sufficient conditions. But to understand these we need to go back to Russell's other objection to Meinong, the one based on the law of non-contradiction. In *Über Möglichkeit und Wahrscheinlichkeit* Meinong revisited his initial reply to that objection as well. Initially he had said simply that the law of non-contradiction doesn't apply to inconsistent objects ([1907], pp. 14-15), eight years later he modifies this response in order to save some form of the law of non-contradiction across the whole range of objects. He makes a distinction between the absence of a property, which he calls the not-being-of-a-so-being (*Nichtsein eines Soseins*), and the presence of a negated property, a not-so-being (*Nichtsosein*). The round square does not both have and lack the property of being round, it has both the property of being round and also the property of being not round ([1915], pp. 171-4). This amounts to a distinction between negating the sentence — it is not the case that the round square has the property of being round — and negating the predicate — the round square has the property of being not-round. The law of non-contradiction does not hold for predicate negation: in general, impossibilia have both a property and its negation, so the round square is both round and not-round. But this does not imply either that some sentence is both true and false, or that a sentence and its negation are both true. The law of non-contradiction does hold for sentential negation. On Meinong's 1915 theory, it is true that the round square has the property of being round, and it is false that it does not have that property; it is also true that it has the property of being not-round, and false that it does not have that property. Lest the sentential inconsistency be revived by simply re-characterizing a round square as both having and lacking the property of roundness, it has to be assumed that objects cannot be characterized by sentential negation.[22] This distinction between sentential and predicate negation has commended itself to many Meinongians, including Jacquette.[23] To avoid confusing the two types of negation, I shall use '$\sim$' for sentential and '$-$' for predicate negation. While '$\sim$' is entirely classical, '$-$', which I shall call 'Meinongian negation', is not very clearly understood.[24] The usual classical equivalence, $\sim Fa \equiv -Fa$, of course, fails in general, and so, too, do $Fa \vee -Fa, \sim (Fa \wedge -Fa)$. Double predicate negation (cf. Routley [1980], p. 193) and at least some forms of predicate-negation contraposition are preserved.

---

[22] This is, presumably, also the case where inconsistency is not involved. The colourless chiliagon is not, contrary to what one might naturally have supposed, a chiliagon which lacks the property of being coloured, but one which has the property of being not-coloured.

[23] Cf. Jacquette [1995], p. 30n; [2001], p. 84; [2001a], p. 63; Routley [1980], pp. 88-91, 192-7, 292-3; Parsons [1980], pp. 19-20, 105-6. Instead of 'predicate negation' and 'sentential negation' Jacquette speaks of 'internal negation' and 'external negation' or 'predicate complementation' and 'propositional negation'. See Findlay [1933], pp. 159-62 for an exposition of Meinong's account.

[24] See Griffin [2008], Jorgensen [2004], and Swanson [2011], Ch. 2 for criticism.

Jacquette ([2001], pp. 97-8) uses the distinction between sentential and predicate negation to provide a formal distinction between nuclear and extra-nuclear properties. While $\sim Fa \equiv -Fa$ does not hold generally, Jacquette argues that it does hold in the case of extra-nuclear properties. Consider existence. To say that the golden mountain has the negative property of non-existence, Jacquette argues, is just to say that it is not the case that the golden mountain exists; and to say that it's not the case that the Statue of Liberty has the property of non-existence is just to say that it has the property of existence. Similarly with possibility. To say that the round square has the negative property of being impossible is just to say that it's not the case that the round square is possible. In the case of an extra-nuclear property F, so Jacquette argues, if $\sim Fa$ then $-Fa$, and conversely. For extra-nuclear properties, the logic of predicate negation is entirely classical: the laws of excluded middle and of non-contradiction will both hold. But for nuclear properties, predicate negation is Meinongian. The round square's having the property of being not round does not imply that it does not have the property of being round; whereas it's having the property of being impossible (or of being non-existent) does imply that it lacks the property of being possible (or of being existent). Jacquette thus marks the distinction between nuclear and extra-nuclear predicates by the following two principles (Jacquette, [2001], p. 97):

(N) $\quad \sim (\forall x_1, \ldots, \forall x_n, \forall F^n)(\sim F^n x_1, \ldots, x_n \equiv -F^n x_1, \ldots, x_n)$[25]

(XN) $(\forall x_1, \ldots, \forall x_n, \forall F^n!)(\sim F^n! x_1, \ldots, x_n \equiv -F^n! x_1, \ldots, x_n)$.

(I've altered his notation for predicate negation to my own, which is easier to type. The quantifiers, '$\forall$' and '$\exists$', should of course be understood as ontologically neutral, ranging over the whole domain of objects, non-existent as well as existent.) Jacquette adopts the shriek ('!') notation, which Russell introduced to mark predicative functions, to indicate extra-nuclear predicates. As a Russell scholar, I find that distracting, but worse, by making distinction between nuclear and extra-nuclear predicates a distinction between styles of variables, it becomes impossible to use the notation to deal with questions of whether a particular predicate is nuclear or not, since it will have to instantiate one style of variable or the other. It seems better to introduce 'is nuclear' and 'is extra-nuclear' explicitly in the formal language as predicates of predicates. For this purpose I use the bold-face predicate letters '**N**' and '**X**'. Thus we can give explicit definitions of 'nuclear' and 'extra-nuclear':

(N*) $\quad \mathbf{N}(F^n)$ iff $\sim (\forall x_1, \ldots, \forall x_n)(\sim F^n x_1, \ldots, x_n \equiv -F^n x_1, \ldots, x_n)$

---

[25]There is a problem with (N). Since the universal predicate quantifier occurs within the scope of the negation, it converts to a particular quantifier in prenex normal form which is plainly not what is intended. The matter is dealt with below.

(XN*) $\mathbf{X}(F^n)$ iff $(\forall x_1, ..., \forall x_n)(\sim F^n x_1, ..., x_n \equiv -F^n x_1, ..., x_n)$.

It may well be that this is correct. It's not implausible to suppose that there is some important connection between the nuclear/extra-nuclear distinction and Meinong's distinction between the two types of negation: one might perhaps think of predicate negation as a watered-down form of sentence negation. Nonetheless, it gives us less help in classifying predicates as nuclear or extra-nuclear than we might have hoped for. Are we really so certain that we know, for any given predicate, whether its predicate negation is classical or not? Are we even sure we know how to find that out? Moreover, even if Jacquette is right that the negation of extra-nuclear predicates is classical and that of nuclear predicates is not, we still don't have any explanation of why this is so. It seems to be some deep feature of the non-existent, but one of which we lack an appropriately deep understanding. Without such an understanding, however, we are not in a position to say which predicates have which sort of negation. Take the predicate 'is complete' and its negation 'is incomplete' and suppose that we try to characterize a non-existent object a as complete. If, as everyone agrees, 'complete' is extra-nuclear, then on Jacquette's proposal 'a is complete' implies that 'a is incomplete' is false. (In a bit more detail: if a is complete, i.e. if Ca, then $--Ca$, by double negation, so by (XN*), $\sim -Ca$, i.e. 'It is not the case that a is incomplete', so 'a is incomplete' is false.) But this relies on the prior information that 'complete' is extra-nuclear. How do we know that we should apply (XN*) here and not (N*)? If we lack that information about completeness we don't have any way of recovering it from independent knowledge of the behaviour of its predicate negation. Predicates in natural language don't come with a marker to indicate whether their negation is classical or Meinongian and I don't think that our intuitions about how the negation of a given predicate works are sufficiently clear to enable us unequivocally to categorize the predicate as either nuclear or extra-nuclear using (N*) and (XN*). In short, our intuitions as to the behaviour of a predicate when it's negated are neither surer nor clearer than our intuitions as to which predicates are nuclear and which extra-nuclear.

Things are not improved if we approach the issue from the other side of the nuclear/extra-nuclear divide. By (N*), a one-place predicate F is nuclear iff $(\exists x)(\sim Fx \not\equiv -Fx)$. So, in principle, we could determine that F was nuclear by finding such an object. If there is such an object it will be a non-existent one, because negative predicates of either type applied to existents behave classically. So this is not a matter of empirical investigation: we need to look to the resources of object theory to see if they provide us with such an object and this will depend upon what assumptions we are able to make. Obviously, if freedom of assumption is unlimited, there will be such an object: we can simply assume an object, a, such that '$\sim Fa$'

is false and '$-Fa$' is true (or vice-versa). On the other hand, if assumption is restricted, then whether there is an object that can instantiate $(\exists x)(\sim Fx \not\equiv -Fx)$ will depend upon whether 'F' is an assumptible predicate or not, i.e., on whether 'F' is nuclear or extra-nuclear — and we are back to where we started. (N*) and (XN*) may very well be correct, and perhaps even important, principles in object theory, but they give us no help in laying out a precise demarcation between nuclear and extra-nuclear predicates.

The absence of a precise demarcation might not matter very much. Not every distinction can be sharply defined in terms of something clearer and more fundamental and Meinongians can point to a fair amount of agreement about particular cases. As Routley says, it is rare that more than this can be offered with any really fundamental distinctions ([1980], p. 264n). What should worry them more is that, by disqualifying extra-nuclear properties from genuinely characterizing objects, they impose limits on freedom of assumption and thereby sin against one of the main tenets of Meinong's theory. That assumption should be absolutely unlimited was Meinong's most radical thought, and the one that led directly to a strikingly simple semantics with good prospects for handling the entire range of discourse — extensional, intentional, factual, fictional, and delusional — in a natural, straight-forward manner without special exemptions, dubious reconstructions, or *ad hoc* devices. Once the nuclear/extra-nuclear distinction is admitted, not only is the straight-forward simplicity of the theory lost, but, with it, quite a bit of its power. No longer are all assumptions what they seem to be: those involving extra-nuclear predicates turn out to be not assumptions in the Meinongian sense at all.

Meinong himself, as we've seen, attempted to mitigate the damage by introducing watered-down nuclear predicates to stand in for extra-nuclear ones, though no one has been able to offer a clear account of what this distinction amounts to. But even if we did have a clear account — and a convincing one — the fact that full-strength extra-nuclear predicates cannot be used to characterize non-existent objects is still a loss to freedom of assumption, even if we have watered-down nuclear stand ins for them. If there is a real difference between a full-strength property and its watered-down surrogate, then there is a real restriction on what assumptions can be made. Jacquette's plausible linking of the nuclear/extra-nuclear distinction with the predicate/sentential negation distinction extends the limitation to assumptions which involve negation. We can make an assumption about a round square which is round and predicatively not round, but we cannot make one about a round square which is round and sententially not round (the strongly round square, we might call it). Yet, if assumption is free, we should be able to make assumptions about the latter as easily as about the former. Indeed, those who were unaware of Meinongian thinking about predicate negation would naturally suppose they had been thinking of the strongly round square all along, for the strongly round square is not some special, esoteric object conjured up to embarrass Meinongians, but the default object when round squares are being discussed.

The same, of course, can be said about full-strength extra-nuclear properties. If we characterize a non-existent object as complete, we would normally think that this was the same full-strength property that we might attribute to an existent object. When it comes to predication, full-strength is surely the norm. It would (or should) come as a surprise to learn that in this we were mistaken, that our assumption could only be that the object had the watered-down version of the property. The only thing that makes this look plausible is the obscurity of the distinction between full-strength and watered-down properties: in the absence of a clear distinction one is apt to assume that there is not much difference between attributing a full-strength property and a watered-down one. But, for the Meinongian theories that invoke the device, the distinction is crucial: on pain of refutation, full-strength extra-nuclear properties are not assumptible. Assumption, therefore, is not free and, in the absence of a better account of watering-down, it is far from clear what the watered-down surrogate properties do to mitigate the situation. Moreover, the obscure full-strength/watered-down distinction is imposed to work in tandem with the equally obscure nuclear/extra-nuclear distinction. It would be theoretically simpler to stipulate that non-entities had only watered-down properties. This would obviate the need for an additional distinction between nuclear and extra-nuclear properties, but at the cost of imposing a global restriction on free assumption instead of a piecemeal one. Alternatively, as Jacquette and Routley both propose, one could make do with only the distinction between nuclear and extra-nuclear predicates and accept the resulting piecemeal restrictions on free assumption without the benefits of mitigation — since these last are dubious to say the least. The theoretical upshot seems to be this: if we could clarify either the distinction between full-strength and watered-down properties or that between nuclear and extra-nuclear properties and justify it on general object-theoretic grounds, and not just as an *ad hoc* device to avoid counter-examples, we could dispense with the other distinction. As things stand, however, neither distinction has been either clarified or adequately justified.

A way around this impasse is offered by the Mally-Zalta approach which replaces the Meinong-Jacquette distinction between different types of predicate by a distinction between different modes of predication, encoding and exemplification (to use Zalta's terms). Existent objects only exemplify properties and never encode them, non-existents both encode and exemplify properties. For example, the existent golden mountain encodes, but does not exemplify, the properties of being golden, being a mountain, and existing; it exemplifies, but does not encode, the properties of being non-existent and being possible. The non-existent golden mountain, on the other hand, both encodes and exemplifies the property of non-existence. Objects may be encoded inconsistently but they may not exemplify inconsistent properties. Identity principles are different for existents and non-existents: existents are identical if they exemplify the same properties, non-existents if they encode the same properties. Accordingly, the non-existent golden mountain is a different object from the golden mountain simpliciter.[26]

Kit Fine ([1984], p. 98) has suggested that Jacquette's theory and Zalta's are inter-translatable, which seems plausible to me (though Zalta [1992] has rejected it). Jacquette claims that the nuclear/extra-nuclear distinction is more fundamental, and attempts to show that the encoding/exemplifying distinction can be reduced to it (Jacquette [1989] and again in [1997]).[27] I cannot see that either doctrine is more fundamental than the other, but it does seem to me that the two modes position is clearer, if only because it doesn't require us to partition an infinite set of properties into two groups — in effect, instance by instance. By contrast, on the two modes account any predicate whatsoever can be encoded or exemplified so we have only to decide of any given attribution of a predicate to a non-existent object whether it is an encoding attribution or an exemplifying one. Jacquette rightly complains that Zalta gives no general principle by which this can be determined (Jacquette [1997], p. 251; cf. Zalta [1988], pp. 30-1) — though, in this respect, I think Zalta's theory fares no worse than Jacquette's own. What seems to decide that an ascription of F to a non-existent object a is encoding is that being F is part of the characterization of a. But this won't do, for then Zalta's distinction between encoding and exemplifying predications would depend upon Jacquette's distinction between nuclear (characterizing) and extra-nuclear (non-characterizing) predicates. So maybe a encodes F only when a is actually described as being F, i.e., when 'a' is a definite description which contains 'F'. This, of course, does not help with non-existent objects that are introduced by means of proper names, even on a description theory of names unless we can provide each name with a canonical description which incorporates exactly its encoding properties. Moreover, it is not sufficient to fix the distinction, since a non-existent object may both encode and exemplify F. So we need also to specify when the object exemplifies F. Here the temptation — which is unavoidable until some clear alternative is spelled out — is to say that it exemplifies F when it really is F. But then a encodes F when it is described as being F,

---

[26]The main source for this view is Zalta [1983]; but see also Rapaport [1978] who takes a similar line.

[27]Jacquette ([1997], p. 250) also complains that Zalta's theory is not really Meinongian at all because Zalta's non-existent objects (Zalta calls them 'abstract objects') all have being. In this respect Zalta's theory is more Platonic than Meinongian. For present purposes only, I shall treat this as an inessential feature of Zalta's position.

and a exemplifies F when it really is F: the non-existent golden mountain encodes non-existence because that is how it is described; but it exemplifies non-existence because it really doesn't exist. This seems to me the most serious problem with the Mally-Zalta approach. It makes it difficult to avoid concluding that the exemplified properties are the only ones a non-existent object really has and that its encoded properties are those it is merely described, or thought of, as having.

One advantage that Zalta's theory has over Jacquette's lies in the identity conditions it supplies for non-existent objects: on Zalta's theory, as we have seen, non-entities a and b are identical if they encode the same properties. On Jacquette's theory they are identical if they have the same nuclear properties.[28] Since any property may be encoded, this allows Zalta to distinguish objects which Jacquette is forced to identify, e.g., the object picked out by 'the existent golden mountain' and that picked out by 'the golden mountain'. But 'the existent golden mountain' is a perfectly good description, so Jacquette is forced to suppose that 'existent' is a kind of vacuous element within it, an element which can be ignored since it makes no contribution to determining the description's reference. Such a view is hard to defend, especially for a Meinongian. It is easy, for example, to imagine a fiction in which the characters embark on a difficult expedition to a hard-to-reach part of their world to try to find a golden mountain which doesn't exist, while they fail to notice a golden mountain which does exist in their own backyard. (It could be a parable on our propensity not to see the value of what exists around us, while we seek it in exotic places where it is not to be found.) Obviously, in order to understand the story, we have to recognize that the two descriptions refer to different objects. It is not clear how Jacquette can do this.[29]

Zalta, of course, can readily distinguish the two: they are encoded differently and thus are distinct objects. The cost for Zalta is the primitive and unexplained distinction between encoding and exemplification. Moreover, if assumption is really free, then we should be able to characterize a non-existent object as exemplifying properties. Indeed, to reprise a line of thought already utilized, one would normally assume that, when one characterized an object as being a mountain and being golden, one *was* characterizing it as *exemplifying* these two properties: when it comes to

---

[28]At least, this is the account he gives in a number of works, e.g., Jacquette [1995a], p. 180; [2001], p. 89; [2001a], p. 78; [2015c], pp. 214-5; [2015d], pp. 310, 314; [2015e], p. 382.

[29]Having said this, I discover that Jacquette (in a very late essay) revived the modal moment he had earlier rejected in order to allow appropriately watered-down extra-nuclear properties to characterize objects so as to deal with an example formally analogous to the one I have just given (cf. [2015b], pp. 156-8). I'm not sure if this signalled a late change of direction or not; nor, if it did, how Jacquette proposed to rescue the modal moment from charges he had earlier laid against it. But resurrecting the modal moment is not strictly necessary. As I have just argued, the distinction between nuclear and extra-nuclear predicates can do the job on its own if it is properly deployed.

property-ascription, exemplification is surely the norm. Zalta, however, denies that it is possible to characterize an object as exemplifying a property, for if we could do that we could characterize it as exemplifying a property (like existence) which conflicts with those properties which — it is tempting to say — it *really* exemplifies (like non-existence). But if we say that then it seems that we say that encoding is not a real form of predication at all. And if we say that, then (barring desperate measures) we are back to the classical Russellian view that only existing objects have properties, i.e., the ontological assumption which all the authors discussed in this paper wish to avoid.

There is, in fact, a way to accomplish this without returning either to the ontological assumption or to the modal moment and even without invoking the nuclear/extra-nuclear distinction, but it involves a fairly radical re-working of the logical structure of object theory. It seems to me that what we want to say in connection with my parable of the two golden mountains is that, though neither exists in the actual world, *within the context of the story* the one exemplifies the property of existence and the other exemplifies the property of non-existence. My claim, then, is not that encoding isn't a real form of predication, it looks that way only because encoded properties are not being predicated in the real world. The existent golden mountain really does not exist, not because it merely encodes existence and encoding is not real predication or does not attribute a real property, but because the existent golden mountain does not exist in the real world. In the context of the story it does exist because there it exemplifies, and doesn't merely encode, existence. In fact, there is no need for Zalta's distinction between two modes of predication; what is needed is a distinction between whether the predicational claim is being assessed in the real world or in a story. At the same time, there is also no need to suppose, with Meinong and Jacquette, that there are different types of predicate, nuclear and extra-nuclear. In the context of the story, an object has *all* the predicates by which it is characterized and that includes those that we have hitherto been treating as extra-nuclear. Moreover, it has them full-strength, rather than watered down. Again, there is no need to introduce the watering-down device at all. In fact, we can avoid all three of the distinctions which have given Meinong and the neo-Meinongians so much trouble. All three distinctions were introduced into the theory of objects because (UFA) and (UCP) generate truth-claims about non-existent objects which are known to be false in reality. All three can be avoided if the truth-claims made by object theory are assessed in the fictional context of the story rather than in the real world. The difference lies not in the content or the nature of the truth-claims (i.e., whether the predicate is nuclear or extra-nuclear, or whether the predication is encoding or exemplifying) but in the context in which its truth-value is assessed. In short, we need distinguish between what is actually the case in the real world and what is actually the case in the context of the story.

But not all objects are accompanied by stories. Since by freedom of assumption any concatenation of predicates in a description characterizes an object, most of Meinong's non-existent objects come with little background, they are isolated[30] and radically incomplete objects which have only the properties mentioned in their descriptions (typically, but not necessarily, closed under inferences of some kind). These are the examples of logic books, like Russell's present king of France and Meinong's round square and, of course, the golden mountain itself; and their consideration has largely driven the development of the theory of objects.[31] When one introduces an example, like the present king of France, in order to prove a point or make an argument, one is supposing such an object as one has described. One is not necessarily supposing it to exist nor even to be capable of existence (though one might equally well be supposing that — supposition, after all, is free). The object about which one is making one's supposition has exactly the properties one supposes it to have. If it didn't, it would be the object of a different supposition. Moreover, it has them (unless one supposes to the contrary) in a perfectly ordinary way: weakening, either of the predicate or of the mode of predication, is to be avoided unless explicitly called for by the supposition in question. All this can be accommodated if we extend the notion of context beyond stories, so that each isolated object is treated as belonging to a context of its own, albeit a very thin, impoverished one, which typically lacks other objects beyond the one supposed, and where the object supposed lacks other properties beyond the ones it is supposed to have. That is the charm of logic examples: their simplicity makes it easier to focus on whatever point they were introduced to make. The idea is that for every supposition, whether elaborate (as in a novel) or sparse (as in a logic example), there is what I call a 'context of supposition' (Griffin [2008]) in which things actually are exactly as the supposition supposes.

The proper treatment, then, of isolated objects simply extends the treatment accorded fictional objects. In the context of supposition appropriate to the object, the object has — in a full, un-watered-down, exemplificatory way — exactly those properties — nuclear and extra-nuclear — which are used to characterize it. Where c is the context, a the object and 'F' the predicate, 'Fa' is true in c; more formally:

---

[30]This term comes from Sylvan [1995].

[31]It is not only in logic that such suppositions occur: they are to be found in all branches of theoretical science, e.g., in classical dynamics: 'Let the body T...', as Newton says in *Principia*, Bk. I, Prop. LIII, Problem XXXV; and similarly, many other authors in many other places. It is not to be supposed that the body does or even could exist, for the conditions placed upon it may be impossible of fulfilment (the proof which follows the supposition may be a *reductio*), but it has to be supposed that the body T has exactly the properties Newton goes on to specify.

$V(Fa, c) = \top$. From this is does not follow that 'Fa' is true simpliciter or true in the actual world, @ (i.e., that $V(Fa, @) = \top$); but equally it does not follow that 'Fa' is not really true in c, nor that a is not really 'F' in c (but merely encoded as such), nor that it is not really 'Fa' that is true in c (but only 'ωFa', where 'ω' is Parsons' watering-down operator). Theories of objects of the type so far considered, which I shall call classical object theories, deal only in terms of truth simpliciter. They sought a single framework — the actual world with a vastly augmented domain of (existent and non-existent) objects[32] — on which every sentence of the theory would be true. But, assumption being free and the characterization postulate being unlimited, the theory delivers every possible sentence! Hence the need in classical theories drastically to curtail assumption and restrict the characterization postulate — quite how drastic the needed restrictions were was effectively hidden by the imprecision with which the required restrictions (like the distinction between nuclear and extra-nuclear predicates) were drawn and the variety of obscure devices (like watering-down and predicate negation) used to mitigate them.[33] Classical object theories, thus face a dilemma. In order to treat intentionality adequately — to capture, that is, the full range of what may be imagined and thought about — object theory needs to be immensely powerful, to be capable of generating every imaginable claim about every imaginable object. On the other hand, any attempt to capture this plenitude semantically within a single framework, essentially with a single assignment of truth-values, is doomed to failure. In the first place, any such framework would be incoherent; in fact, utterly trivial, making every possible claim true. But, secondly, even such an accommodating framework would fail to do justice to the semantics of supposition. For although, since assumption is free, anything might be supposed to be true, it is not (normally) supposed that everything is true. Suppositions, in general, are not only much more limited than that but very often involve scenarios which are entirely coherent or, at the very least, are plausible

---

[32]This is the so-called "factual" or "absolute" framework T of Routley's theory of items (cf. Routley [1980], p. 166). It is the extended domain that makes classical object-theories preferable to the traditional, referentialist theories in treating fictions, abstractions, and intentionality.

[33]In this paper, I have not even considered the most egregious of the problems classical item theory faces. These arise in connection with relations, when nonentities are characterized by relations to objects which exist in the actual world (cf. Woods [1969], [1974]). The problem of relations, perhaps the most difficult that classical object theory has to face, has led object theorists to truly desperate expedients (e.g. Routley ([1980], pp. 267–9, 577–90; Parsons ([1980], pp. 26–7, 59–60, 75–7), none of which seems to me remotely satisfactory. I have dealt with relations in some detail in Griffin [2008] and [forthcoming]. The difficulty affects even Russell's favourite example, the present king of France. By (UCP), the present king of France is presently king of France, but France is presently a republic and has no king. This is no accident since Russell's interest in definite descriptions arose from the need to define ordinary mathematical functions (descriptive functions, as he called them to distinguish them from propositional functions) in terms of relations (cf. PM, *30).

contenders for coherence. The only way out of this impasse, without imposing awkward and ultimately crippling restrictions on the theory's main principles, is to distribute the semantic load, in particular the assignment of truth-values, across a variety of frameworks, here called contexts of supposition. Theories of this type I call 'distributive item theories'.[34]

Informally, the semantics for distributive item theory requires a set of contexts **C** and a domain of objects $\mathbf{D}_c$ for each $c \in C$. The interpretation function, **I**, assigns an extension, $\mathbf{I}^+(F)$, and an anti-extension, $\mathbf{I}^-(F)$, to each n-place predicate 'F' in each context, c. Where 'F' is an n-place predicate, $\mathbf{I}^+(F,c)$ and $\mathbf{I}^-(F,c)$ are n-tuples of elements of $\mathbf{D}_c$. The valuation function, V, assigns truth-values in the usual way:

$$V(Fx_1, ..., x_n, c) = \top \text{ iff } \langle x_1, ..., x_n \rangle \in \mathbf{I}^+(F, c)$$
$$V(Fx_1, ..., x_n, c) = \bot \text{ iff } \langle x_1, ..., x_n \rangle \in \mathbf{I}^-(F, c)$$

However, the extension and anti-extension of a predicate at a context cannot be supposed to be exhaustive, since objects (like Mally's parallelogram-*in-abstracto*) may be incomplete with respect to some predicates at that context, nor can they be assumed to be exclusive, since objects (like Meinong's round square) may be inconsistent with respect to some predicates at that context.[35]

Although distributive object theory requires only one kind of predicate and only one mode of predication, it can, I think, make sense of the intuitions that led Meinong and Jacquette and Mally and Zalta, respectively, to suppose otherwise. When trying to expound Zalta's distinction between encoding and exemplification I found it difficult to avoid the conclusion that an object encodes those properties it was thought to have (or was described as having) while it exemplified those it really did have. But that could not be to Zalta's purpose, because being thought to have a property is not a special way of having it. So pursuing object theory in this way leads straight back to the ontological assumption of the referentialist theories, that only entities (really) have properties. On a distributive theory of objects, we do not need to suppose that encoding is a special mode of predication: the existing golden mountain exemplifies all three of the properties used to characterize it, but it exemplifies them in the context of supposition into which it is introduced. In that context it really has those properties. In the actual world it exemplifies the property of not existing. Here, however, exemplification should not be thought of as a contrast to encoding: exemplify is just what objects do when they have properties. Instead of

---

[34] The term 'item' come from Routley [1980], who chose it as being even more inclusive than Meinong's 'object'. The adjective 'distributive' was suggested by Nancy Doubleday. I have defended distributive item theories in Griffin [2008] and [forthcoming]. The feasibility of the distributive approach has been formally demonstrated very elegantly by Priest [2005].

[35] Formal details are given in Priest [2005], though he uses a fixed domain semantics.

Zalta's distinction between encoding and exemplifying as two distinct ways of having properties we have a rather more familiar distinction between having properties in some context of supposition and having them in the actual world.

The same is true of the nuclear/extra-nuclear distinction. On distributive object theory, an object has all the properties by which it is characterized, whether nuclear or extra-nuclear, in some context or other. In the context of supposition, the characterization postulate reigns supreme: which truth-value is assigned to a proposition in a context of supposition depends entirely upon the nature of the supposition, upon how the objects which the proposition is about are characterized. But this is not the case in the actual world. Different views of what properties non-entities have in the actual world may be taken within the general rubric of distributive item theory. One could take an orthodox Meinongian line and assign the existent golden mountain to the extension of 'is golden' in the actual world, thus making it true in the actual world that the existing golden mountain is golden. Alternatively one might take a more referentialist approach and assign the golden mountain to neither the extension nor the anti-extension of 'is golden' in the actual world, thus leaving 'The existing golden mountain is golden' without a truth-value in the actual world. I shall not try to decide this issue here, though I incline to the referentialist option. But on both approaches we have to assign the existing golden mountain to the anti-extension of 'exists' in the actual world, for in the actual world 'The existing golden mountain exists' is false. In the actual world, on either view, characterization has its limits, and those limits, I want to maintain, fall in part on what has been taken to be the demarcation line between nuclear and extra-nuclear predicates. It is not that extra-nuclear predicates are a mysterious, special kind of predicate which cannot be genuinely used, in a full, un-watered-down, exemplificatory way, to characterize non-entities. It is rather the simple, elementary fact (almost always cited by Meinongians as a justification for the nuclear/extra-nuclear distinction) that for some predicates we cannot decide by fiat that a non-existing object falls into their extension *in the actual world*. It is plausible to read Meinong as groping for something like this view when he first introduced the modal moment. '[I]t is beyond my power', he said 'to genuinely reach an *actually* existing object by deliberating upon its nature. My authority inevitably lags behind at exactly the point of the modal moment' ([1915], p. 282; my italics). We cannot reach an *actually* existing object by deliberating on its nature because an actually existing object is not an object that exists in some special way, but just an object that exists in the actual world. And we cannot reach such an object, not because of some limitation of mental or linguistic capacity, but because there may be no such object there to be reached. But, in an appropriate context of supposition, we can reach an object which exists there simply by supposing such a thing. Actual existence is not some special, souped up kind of existence, it is just existence in the actual world. It is not implausible, with a bit of charity, to see Meinong's account of the modal moment as an attempt to make exactly this point, but an attempt muddled by Meinong's presentation of what is essentially a modal or a semantical marker as if it were a predicate modifier.

References

[1] Albertazzi, Liliana, Dale Jacquette and Roberto Poli (eds.) [2001] *The School of Alexius Meinong* (Aldershot: Ashgate).

[2] Allison, Henry [1983], *Kant's Transcendental Idealism*, (New Haven, Conn.: Yale University Press)

[3] Campbell, Richard [1974], 'Real Predicates and "Exists"', *Mind*, 83: 95-99.

[4] Chisholm, Roderick M. [1982], 'Converse Intentional Properties', *Journal of Philosophy*, 79: 537-45.

[5] Findlay, J.N. [1933], *Meinong's Theory of Objects and Values* (Oxford: Clarendon; 2nd edn., 1963).

[6] Fine, Kit [1984], Critical Review of Parsons [1980], *Philosophical Studies*, 45: 95-142.

[7] Griffin, Nicholas [2008], 'Rethinking Item Theory', in Nicholas Griffin and Dale Jacquette (eds.) *Russell vs Meinong. The Legacy of "On Denoting"*, (London, New York: Routledge), pp. 204—32.

[8] Griffin, Nicholas [forthcoming] 'Why Item Theory Doesn't (Quite) Go Far Enough', in Richard Sylvan, *Exploring Meinong's Jungle and Beyond*, corrected and revised edn., ed. by Maureen Eckert, Dominic Hyde, *et al.* (Berlin: Springer)

[9] Jacquette, Dale [1985], 'Meinong's Doctrine of the Modal Moment', *Grazer Philosophische Studien*, 25/26: 423-438.

[10] Jacquette, Dale [1989], 'Mally's Heresy and the Logic of Meinong's Object Theory', *History and Philosophy of Logic*, 10: 1-14.

[11] Jacquette, Dale [1995], 'Defoliating Meinong's Jungle', *Axiomathes*, 1-2: 17-42.

[12] Jacquette, Dale 1995a] 'Meinong's Theory of Implexive Being and Non-Being', *Grazer Philosophische Studien*, 50: 233-71; reprinted in Jacquette [2015a], pp. 163-191

[13] Jacquette, Dale [1996], *Meinongian Logic. The Semantics of Existence and Nonexistence* (Berlin: de Gruyter)

[14] Jacquette, Dale [1997], 'Reflections on Mally's Heresy', *Axiomathes*, 8: 163-80. reprinted in Jacquette [2015a], pp. 247-261

[15] Jacquette, Dale [2000] 'Confessions of a Meinongian Logician'. *Grazer Philosophische Studien*, 58-59: 151-180; reprinted in Jacquette [2015a], pp. 363-380.

[16] Jacquette, Dale [2001], 'Nuclear and Extranuclear Properties', in Albertazzi, *et. al.*, (eds.) [2001], pp. 397-426; reprinted in Jacquette [2015a], pp. 83-109.

[17] Jacquette, Dale [2001a], 'Aussersein of the Pure Object' in Albertazzi, *et. al.*, (eds.) [2001], pp. 373-396; reprinted in Jacquette [2015a], pp. 59-81.

[18] Jacquette, Dale [2008], 'Object Theory Logic and Mathematics: Two Essays by Ernst Mally', *History and Philosophy of Logic*, 29: 167-82; reprinted in Jacquette [2015a], pp. 389-404.

[19] Jacquette, Dale [2009], 'Meditations on Meinong's Golden Mountain', in Nicholas Griffin and Bernard Linsky (eds.) *Russell vs Meinong: The Legacy of 'On Denoting'*, (London: Routledge); reprinted in Jacquette [2015a], pp. 111-43.

[20] Jacquette, Dale [2015a] *Alexius Meinong, The Shepherd of Non-Being* (Berlin: Springer)

[21] Jacquette, Dale [2015b], 'Domain Comprehension in Meinongian Object Theory', in Leclerq et al., [2015], pp. 101-22; reprinted in Jacquette [2015a], pp. 145-161

[22] Jacquette, Dale [2015c] 'About Nothing', first published in Jacquette [2015a], pp. 193-228.

[23] Jacquette, Dale [2015d] 'Anti-Meinongian Actualist Meaning of Fiction in Kripke's 1973 John Locke Lectures', first published in Jacquette [2015a], pp. 301-28.

[24] Jacquette, Dale [2015e] 'Meinongian Dark Ages and Renaissance', first published in [Jacquette [2015a], pp. 381- 87

[25] Jorgensen, Andrew Kenneth [2004], 'Types of Negation in Logical Reconstructions of Meinong', *Grazer Philosophische Studien*, 67, pp. 21—36.

[26] Kant, Immanuel, [1781] *Critique of Pure Reason*, transl. by Norman Kemp Smith (New York: St Martin's Press, 1965; 1st edn. 1929)

[27] Kemp Smith, Norman [1923], *A Commentary to Kant's Critique of Pure Reason*. (Basingstoke: Palgrave; 2nd edn., 2003)

[28] Leclerq, Bruno, Sébastien Richard and Denis Seron (eds.), [2015], *Objects and Pseudo-Objects: Ontological Deserts and Jungles from Brentano to Carnap* (Berlin: de Gruyter)

[29] Locke, John [1690], *An Essay Concerning Human Understanding*, ed. by P.H. Nidditch, (Oxford: Clarendon Press, 1975)

[30] Mally, Ernst [1908], 'Gegenstandstheorie und Mathematik', *Bericht über den III. Internationalen Kongress für Philosophie zu Heidelberg*; English translation in Jacquette [2008].

[31] Mally, Ernst [1912], *Gegenstandstheoretische Grundlagen der Logik und Logistik* (Leipzig: Barth).

[32] Meinong, Alexius [1904], 'The Theory of Objects', transl. I. Levi, D.B. Terrell, and R.M. Chisholm in Chisholm (ed.) *Realism and the Background of Phenomenology* (Glencoe, Ill.: The Free Press, 1960), pp. 76-117.

[33] Meinong, Alexius (ed.) [1904a], *Untersuchungen zur Gegenstandstheorie und Psychologie* (Leipzig: Barth).

[34] Meinong, Alexius [1907], *Über die Stellung im System der Wissenschaften* (Leipzig: Voigtländer).

[35] Meinong, Alexius [1910], *On Assumptions* (2nd edn.; 1st edn. 1902), transl. by James Heanue, (Berkeley, Calif.: University of California Press, 1983).

[36] Meinong, Alexius [1915], *Über Möglichkeit und Wahrscheinlichkeit* (Leipzig: Barth).

[37] Parsons, Terence [1980], *Nonexistent Objects* (New Haven: Yale University Press)

[38] Priest, Graham [2005], *Towards Non-Being. The Logic and Metaphysics of Intentionality* (Oxford: Oxford University Press; 2nd edn. much expanded, 2016).

[39] Rapaport, William J. [1978], 'Meinongian Theories and a Russellian Paradox', *Noûs*, 12: 153-80.

[40] Routley, Richard [1980], *Exploring Meinong's Jungle and Beyond* (Canberra: Research School of Social Sciences, Australian National University).

[41] Russell, Bertrand [1905], 'On Denoting' in Russell [1994], pp. 415-427.

[42] Russell, Bertrand [1905a], Review of Meinong (ed.) [1904a], in Russell [1994], pp. 596-604.

[43] Russell, Bertrand [1907], Review of Meinong [1907], in Russell, [2014], pp. 689-92.

[44] Russell, Bertrand [1910], 'Knowledge by Acquaintance and Knowledge by Description' in Russell [1992], pp. 148- 161

[45] Russell, Bertrand [1992] *The Collected Papers of Bertrand Russell, vol. 6, Logical and Philosophical Papers. 1909-13*, ed. by John G. Slater (London: Routledge)

[46] Russell, Bertrand [1994], *The Collected Papers of Bertrand Russell, vol. 4, Foundations of Logic 1903-05*, ed. by Alasdair Urquhart (London: Routledge).

[47] Russell, Bertrand [2014], *The Collected Papers of Bertrand Russell, vol. 5, Toward "Principia Mathematica", 1905-08*. ed. by Gregory H. Moore, (London: Routledge)

[48] Swanson, Carolyn [2011], *Reburial of Nonexistents. Reconsidering the Russell—Meinong Debate* (Amsterdam: Rodopi).

[49] Sylvan, Richard [1995], "Re-Exploring Item Theory. Object Theory Liberalized, Pluralized and Simplified but Comprehensivized", *Grazer Philosophische Studien*, 50, pp. 47—85.

[50] Whitehead, A.N. and Russell, Bertrand [1910-13] *Principia Mathematica* (Cambridge: Cambridge University Press; 2nd edition, 1925-7), 3 vols.

[51] Wood, Allen W. [1992], 'Rational Theology, Moral Faith, and Religion', in Paul Guyer (ed.) *The Cambridge Companion to Kant*, (Cambridge: Cambridge University Press), pp. 394-416.

[52] Woods, John [1969], "Fictionality, and the Logic of Relations", *Southern Journal of Philosophy*, 7: 51—63.

[53] Woods, John [1974], *The Logic of Fiction* (The Hague: Mouton).

[54] Zalta, Edward N. [1983], *Abstract Objects. An Introduction to Axiomatic Metaphysics* (Dordrecht, Kluwer).

[55] Zalta, Edward N. [1988], *Intensional Logic and the Metaphysics of Intentionality* (Cambridge, MA.: MIT Press)

[56] Zalta, Edward N. [1992], 'On Mally's Alleged Heresy: A Reply', *History and Philosophy of Logic*, 13: 59-68.

On the Existential Import of General Terms

Guido Imaguire
Universidade Federal do Rio de Janeiro (UFRJ), Brazil.
guido_imaguire@yahoo.com

Abstract

In his book *Logic and How It Gets That Way* Jacquette (2010) presents 'the formalization paradox' which emerges from the attempt to formalize a sentence like 'some monkey devours every craisins', where craisins are imaginary non-existent fruits. From this paradox Jacquette concludes the expressive inadequacy of classical predicate quantificational logic. In this paper I analyse the three assumptions supposed in the emergence of the paradox, viz.: (i) colloquial expressions of the same logical form can and should be formally symbolized by applying the same symbolization schema); (ii) 'Some monkeys devours every raisin' is correctly translated as $\exists x[Mx \wedge \forall y[Ry \to Dxy]]$; (iii) uninstantiated predicates can legitimately enter into (meaningful, true or false) predicates-quantificational symbolizations. I fully accept (iii), but reject both (i) and (ii). I argue, firstly, that (i) has at the first glance two possible interpretations, a trivial and a false one. So, I try to establish a third and more reasonable interpretation. Based on this interpretation I argue that $\exists x[Mx \wedge \forall y[Ry \to Dxy]]$ is not the adequate formalization of 'some monkeys devours every raisin'. My basic claim is based on a generalisation of Russell's theory of description: just like most sentences of natural language which contain definite descriptions are viewed as entailing existential force which must be made explicit in formalization, so do we also consider many, although not all, sentences which contain general terms. A criterion for deciding in each possible case if the sentence entails existential force will be presented and defended.

In his book *Logic and How It Gets That Way* (2010, chap. 2) Jacquette presents 'the formalization paradox' which is supposed to emerge from the attempt to formalize some colloquial sentences into classical predicate quantificational logic. Based on this paradox Jacquette concludes the expressive inadequacy of this logic.

I would like to thank to the audiences of the POL Colloquium in Curitiba, the Sixth International Symposium on Philosophy of Language and Metaphysics att UFF and the LanCog Group in Lisbon, in particular to Breno Hax, Dirk Greimann, and Ricardo Santos for many important suggestions.

In this paper I analyse Jacquette's supposed paradox. Firstly, I will reject that there is a strict paradox at all. What we must recognize instead is rather the expressive inadequacy of standard formalization strategy. Secondly, I will analyse the three assumptions that, according to him, are supposed in this strategy. In particular, I will discuss the 'assumption of logical formalism': colloquial expressions of the same logical form can and should be formally symbolized by applying the same symbolization schema. I will argue that in order to capture in classical logic the intended meaning of colloquial sentences we must take into account the existential force of some sentences. From this, I will propose a new and, as I hope, more adequate formalization strategy.

1 The Formalization Paradox

According to Jacquette, a paradox emerges when we try to formalize some colloquial sentences into classical predicate logic. He observes correctly that a sentence like

(S1) Some monkey devours every raisin.

should not be translated as

(T0) $\exists x \forall y [[Mx \land Ry] \to Dxy]$

since (T0) is not committed to the existence of any monkeys, while (S1) is. As most logicians, Jacquette suggests that a satisfactory translation of (S1) in classical framework would be something like (T1):

(T1) $\exists x [Mx \land \forall y [Ry \to Dxy]]$

However, if the translation of (S1) into (T1) is correct then we ought to apply the same symbolization schema to another sentence involving a make-up type of non-existent fruit, called 'craisins'. Since there are no craisins, it sounds false to say that some monkey devours every craisin. Although there are monkeys, 'none of these existent monkeys devours any let alone every craisin, since craisins there are none' (p. 25). Therefore, it is intuitively false that

(S2) Some monkey devours every craisin.

Anyway, following the same schema for the general form $R(some\ F,\ every\ G)$, we should formalize (S2) as

(T2) $\exists x [Mx \land \forall y [Cy \to Dxy]]$

Now, since by hypothesis there are no craisins ($\neg \exists x Cx$), it follows in a classical logic environment that (T2) is true, which clearly contradicts our intuition that (S2) must be false. This suggests that (T2) cannot be an adequate translation of (S2).

Take now the predicate 'K' for imaginary non-existent animals, kmonkeys. By parity of form, a sentence like

(S3) Some kmonkey devours every raisin.

should be translated as

(T3) $\exists x[Kx \wedge \forall y[Ry \to Dxy]]$

Note that both (S3) and (T3) are false for there are no kmonkeys. So far so good. Jacquette's paradox arises at this point. It is intuitively true that

(S4) No monkey devours every craisin.

Following the established symbolization schema, we should translate (S4) as

(T4) $\neg \exists x[Mx \wedge \forall y[Cy \to Dxy]]$

Curiously, Jacquette claims that (T4) is true (sic!). However, since (T4) syntactically contradicts (T2), it follows that (T2) must be false. But this is highly surprising, for we concluded before that (T2) must be classically trivially true on the grounds that there are no craisins.

The first point we must note is that even a brilliant logician like Jacquette is not infallible. In fact, (T4) is not true as he claims, but false. The condition of the existence of monkeys is empirically satisfied and the additional condition '$\forall y[Cy \to Dxy]]$' is vacuously satisfied; therefore, the negative formula is false. Thus, there is no paradox in the strict sense of the word.

However, there is indeed a difficulty in standard methods of formalization: the intuitively false sentence (S2) is translated into the true (T2), while the intuitively true sentence (S4) is translated into the false (T4). In our standard formalization what should be true is false, and what should be false is true. We may call this difficulty 'the expressive inadequacy problem'. In any case, according to Jacquette's diagnosis, the problem of traditional formalization follows from the combination of three reasonable assumptions, viz.:

(i) colloquial expressions of the same logical form (in this case, R[some φ, every ψ]) can and should be formally symbolized by applying the same symbolization schema ('assumption of logical formalism');

(ii) 'Some monkeys devours every raisin' is correctly translated as $\exists x[Mx \wedge \forall y [Ry \rightarrow Dxy]]$;

(iii) uninstantiated predicates can legitimately enter into (meaningful, true or false) predicates-quantificational symbolizations;

Jacquette claims finally that, if there is no prospect of discarding these assumptions, the only remaining alternative is to turn away from classical logic and recognize the need to go non-classical. Here we have four main possibilities: relevance logic, paraconsistent or dialethic logic, free logic and intensional ('Meinongian') logic.

But, is there really no way of solving the expressive inadequacy problem in the framework of classical logic? This is the question to be addressed in the rest of this paper. My claim is that there is no need to go non-classical, at least not for the reason given here. Not classical logic if fault, but our standard strategy of formalization. I will analyse the three assumptions and conclude that: (iii) is correct, (i) is ambiguous and must be qualified and (ii) is false.

2 The Uninstantiated Predicates Assumption

Lets start with the third and, in my view, less problematic assumption:

(iii) uninstantiated predicates can legitimately enter into (meaningful, true or false) predicates-quantificational symbolizations.

Since, as I think, the inadequacy problem does not emerge from this assumption, I will simply accept it without further discussion. As Jacquette (2010:26) correctly argues, 'if we deny that uninstantiated predicates can legitimately enter into predicate-quantificational symbolizations, then we pay a rather unacceptable high price', for we certainly want to say in many occasions that 'there are no Fs' or that 'Fs do not exist'. Indeed, this sounds like an unobjectionable reason.

Could we imagine a possible motivation for rejecting (iii)? I can see only one reason, namely, for reasons of parity: usually, we do not allow the use of a vacuous individual constant in logic. There is in logic no constant c, such that 'c' is the name of nothing. So, why should we accept uninstantiated predicates?

However, there is a quite substantial reason for a different treatment of individual constants and predicates, viz. set theory which constitutes the standard semantics for predicate logic. While there is an obvious semantic value for uninstantiated predicates, viz. the empty set, there is no possible value for a vacuous individual constant. It would be plain non-sense to claim that the semantic value of the empty constant 'α' is that non-existing member of the empty set.

In fact, the uninstantiated predicates assumption is only interesting in connection with the assumption (i), 'the assumption of logical formalism'. For, once we allow the use of uninstantiated predicates in symbolizations, a second and more interesting question arises, namely, do these predicates enter into formalization in the same way as their instantiated counterparts? With other words, two colloquial sentences of the same logical form, one with an instantiated and the other with a uninstantiated predicate, can and should be formally symbolized by applying the same symbolization schema? In fact, I even think that there are some reasons for accepting a strengthened version of (iii): uninstantiated predicates can legitimately enter into predicates-quantificational symbolizations and by applying exactly the same symbolization schema. This is a plausible (although not necessary) consequence of jointly accepting (i) and (iii). So, lets examine (i).

3 The Assumption of Logical Formalism

Jacquette calls (i) the 'assumption of logical formalism'. So, let's refer to it as '(ALF)':

(ALF) Colloquial expressions of the same logical form (in this case, $R[\text{some } \varphi, \text{ every } \psi]$) can and should be formally symbolized by applying the same symbolization schema.

(ALF) tells us something about 'colloquial expressions' in general, including apparently whole sentences and sub-sentential expressions as well. Thus, we must take both into account. Lets start with sentences.

At first glance, (ALF) seems to have two possible readings, a trivial and a false one depending on how we understand the expression 'logical form'. According to the trivial reading, sentences with the same 'deep' logical form should be symbolized by applying the same symbolization schema. This reading strikes me as trivial and, as far as I can see, no one would dare reject it. After all, the 'logical form' of a sentence is nothing else as a kind of idealized logical structure represented by a formulae that perfectly codifies its formal properties. This way, to formalize two sentences of the same logical form by different symbolization schemes would be inadequate by definition. Of course, this does not imply that formalization of natural expressions is an easy task, nor that logicians always agree about which is the particular 'deep' logical form of a given colloquial sentence, nor, finally, that there is always one univocal answer.

According to the false reading, sentences of the same 'superficial' (merely grammatical) form should be symbolized by applying the same symbolization schema.

Virtually every logician today will agree that sentences of the same grammatical form may have different logical forms, as Frege's (1884/1950:60) famous example clearly illustrates:

(S5)   The whale is a mammal

(S6)   The number three is a prime

We could perhaps try to establish a third intermediate reading between the trivial and the false one. We may suppose that there is an intermediate level between the deep logical form and the superficial grammatical form. When a logician argues that (S5) should be properly formalized as '$\forall x[Wx \to Mx]$', instead of simply 'Fa', an usual speaker will be perplexed about how a formula that entails the logical notions 'all' and 'if then' can properly capture the meaning of a sentence with no apparent logical term at all. In order to make explicit what is really going on when we formalize a colloquial sentence we could indicate a step-by-step paraphrase procedure for approximating the adequate logical form:

(S5)   The whale is a mammal.

(S5)*   All whales are mammals.

(S5)**   Every whale is a mammal.

(S5)***   If something is a whale, then it is a mammal.

(F5) $\forall x[Wx \to Mx]$.

(S5)* provides a more explicit reading of what is meant by the sentence (S5), and both (S5)** and (S5)*** are 'quasi-formalized' versions of (S5)*. 'Quasi-formalized' sentences are sentences of natural language whose grammatical form is intermediate between the original sentence of natural language (i.e. the most natural way to express a proposition) and the formalized expression. There may be indefinitely many semi-formalized sentences between the original sentence and its logical form, although the number is plausibly in most cases relatively small and depends on the pretended granularity of the analysis. We could try to use the number of stars for signaling in a very loose way the approximation to the formalized expression. However, since there may be different routes to approximate to pure formalized expression of the sentence, the number of stars should not be taken too seriously (e.g. it is not clear that (S5)** is more closely connected to the formalized expression than (S5)*).

Anyway, this may have been what Jacquette had in mind with assumption ALF: colloquial sentences of the same semi-formalized form should be formally symbolized by applying the same symbolization schema. Thus, once we recognize that what we want to express is something like (S5)***, the formalization of any sentence with a quasi-formalized form 'if something is a F, then it is a G' must follow the same pattern. The interesting question becomes than whether (S5)*** is an adequate paraphrase of (S5) or not.

What about the logical form of sub-sentential expressions? Let's restrict ourselves to simply subject-predicate sentences. We may distinguish two cases of (ALF), one for the subject term and one for the predicate:

(ALF-pred) For any sentence of the form subject-predicate, it should not make any difference in the logic form, whether the predicate is instantiated or not.

(ALF-sub) For any sentence of the form subject-predicate, it should not make any difference in the logic form, whether the subject term is vacuous or not.

According to (ALF-pred) 'some monkey devours every raisin' and 'some kmonkey devours every craisin' have the same logical form; and according to (ALF-sub) 'Bertrand Russell is British' and 'Sherlock Holmes is British' have the same logical form.

Although Jacquette is concerned with predicates, a brief discussion on (ALF-sub) will lead us to important insights. The issue on how to formalize singular terms of natural language is a highly complex topic that has been exhaustively discussed in analytic philosophy. It involves basically the distinction between proper names and definite descriptions — both may be vacuous or not — and the question how to translate them into logic. There are four main strategies for dealing with both kinds of singular terms in logic.

According to the first strategy, one can and should translate both into simply individual constants like 'a'. In this case if we assume that a singular term of natural language is empty, the best we can do is to accept that it has sense but not reference. According to Frege (1982), sentences containing empty singular terms will not have any truth-value and should be excluded of the scope of logic.

According to the second strategy, we could represent (logical or ordinary) proper names by individual constants like 'a' and definite descriptions, as Russell (1905) suggested, by a quantified expression like

(DD) $\exists x[\phi x \wedge \forall y[\phi y \to x = y]\ldots]$

According to the third strategy, we could completely avoid individual constants in symbolization and translate both proper names and definite descriptions in the

Russellian style. This may sound extreme, but it seems to be well justified if we assume that all ordinary names are truncated descriptions or if we are worried about the ontological import of names of fictional entities like 'Pegasus', as e.g. Quine (1950) was. This way, all sentences of the form proper name-predicate or definite description-predicate, independently of having in the subject position an empty or a non-empty singular term, have exactly the same logical form, viz. (DD). In this case, the referential role of the name is guaranteed by a corresponding adequate predicate, which may be an associated description like 'is the winged horse that was captured by Bellerophon', or a verbalization of the proper name like 'being Pegasus' or 'pegasizes'. Indeed, this is a quite elegant strategy for dealing with singular terms given its ontological parsimony. However, for keeping this option open, we must allow the use of the uninstantiated predicates. Thus, this yields a further argument for accepting the assumption (iii): we need uninstantiated predicates like 'pegazises' in order to avoid the commitment to non-existent fictional entities.

Finally, according to the fourth strategy, we should distinguish the logical form of a definite description in subject and in predicate positions. Based on Donnellan's (1972) distinction between referential and attributive use of definite descriptions, Chateaubriand (2001: chap. 3, and similar in 2002) suggested for instance that a definite description in the subject position may be represented by a constant while the same definite description in the predicate position may be represented by the Russellian quantified form. As a result, the sentences 'Quine is the author of Word and Object' and 'the author of Word and Object is Quine' have different forms.

Be our decision concerning singular terms as it may, for the purpose of this paper, the relevant conclusion is this: despite all differences in the accounts on how to formalize colloquial singular terms into logic they all agree in one point, viz. that all singular terms of the same kind (names or descriptions) and in the same position (or the same kind of use), whether vacuous or not, have the same logical form and should be formalized the same way. Given two singular terms 'α' and 'β', such that one of them is empty and the other not, any two sentences S and $S*$ of the equivalent grammatical form ('β' replaces in $S*$ any occurrence of 'α' in S, i.e. $S* = S[\beta/\alpha]$) must be formalized according to the same pattern. In one word, all strategies accept (ALF-sub).

Before we continue with (ALF-pred), a brief remark: Ironically, as a possible solution for his paradox, Jacquette mentions so-called intensional Meinonguian logic. Zalta's (1983 and 1988) logic is one on them — probably the most developed one. Basically, this logic was developed for regimenting our talking and thinking about domains of non-existing objects. One central feature of this logic is the distinction between the relations of instantiation and codification. While Bertrand Russell really instantiates the property of being British, Sherlock Holmes only codifies this property. For marking this distinction in logic, Zalta introduces the formal distinction between 'Fa' (when a really instantiates F) and 'aF' (in the case a 'only' codifies F). However, it seems to be a clear shortcoming when, given a natural sentence like 'Peter is British' we can only decide its logical form (if it is 'Fa' or 'aF') after we know if 'Peter' in this context is a real or a merely fictional name. Thus, this intensional logic does not obey (ALF-ST). I think that this clearly undermines Jacquette's own suggestion to reject classical logic and replace it by intensional logic. This logic only appears to be more appropriate to formalize the monkey sentence because it is a sentence without singular terms. If (ALF) is a desirable assumption, for dealing with sentences with singular terms, classical logic is still superior than intensional logic.

Now, what about (ALF-pred)? In fact, I think we must accept it. We do not only have to accept that uninstantiated predicates can legitimately enter into (meaningful, true or false) predicates-quantificational symbolizations, but also that they must enter exactly in the same way as their instantiated counterparts. And the reason is the same as for singular terms: the decision about the logical form of a sentence should not depend on additional non-logical knowledge about the actual existence of instances of the predicates. Therefore, the following four sentences must have the same form:

(S1)   Some monkey devours every raisin.

(S2)   Some monkey devours every craisin.

(S3)   Some kmonkey devours every raisin.

(S7)   Some kmonkey devours every craisin.

But this is not all we can say. As we saw above with the whale-prime number example, sometimes colloquial sentences with the same superficial grammatical form have different logical forms. And we can only realize this when we scrutinize more closely the intended meaning of the sentence or utterance. And this, finally, can only be done when we pay attention to some more general contextual and semantical aspects, as I want to argue now.

Take the two sentences

(S8)   *The present King of France is not bald.*

(S9)   *The present King of France does not exist.*

It seems obvious that both sentences, despite their similar grammatical form, have different logical forms. For understanding this better, lets assume Russell's Theory of Descriptions. This is one of the most influential theories on the general topic of logical forms and may serve as inspiration for the rest of our paper. I think that the recipe of its success is due to its sensitivity to grasp the 'deep' intended meaning of a particular kind of sentence. According to the Russellian orthodoxy, a sentence of the form the F is G has the form $\exists x[Fx \land \forall y[Fy \to x = y] \land Gx]$, i.e. it 'entails' or 'is committed to' the existence and uniqueness of 'the F'. Indeed, when we say that the king of France is bald or not bald, we seem to be committing ourselves to the existence of a unique king of France. So, I think most Russellians would agree with the following step-by-step paraphrase procedure

(S8)   *The present King of France is not bald.*

(S8)* *There is exactly one present King of France and he is not bald.*

(S8)** *There is something x that is a present King of France, and for any entity y, if y is a present King of France it is identical to x, and x is not bald.*

(T8) $\exists x[Fx \land \forall y[Fy \to x = y] \land \neg Gx]$

However, certainly no one would suggest a similar step-by-step paraphrase procedure for (9):

(S9)   *The present King of France does not exist.*

(S9)* *There is exactly one present King of France and he does not exist.*

(S9)** *There is something x that is a present King of France, and for any entity y, if y is a present King of France it is identical to x, and x does not exist.*

(T9) $\exists x[Fx \land \forall y[Fy \to x = y] \land \neg \exists x[Fx]]$

which is, of course, non-sense. (S9) is a perfectly reasonable and true sentence, while (T9) is contradictory. In fact, already in step (S9)* of the paraphrase procedure we recognize the non-sense. Although the sentences (S8) and (S9) have the same grammatical form, they do not have the same quasi-formalized form and should not be formalized the same way. The reason for the formalization discrepancy is clear: the (T9) entails an existential commitment that (S9) explicitly does not entail. In fact, I think there are many colloquial sentences besides 'a does not exist' which do not entail such commitment, e.g.

Pegasus is a merely fictional character.
The golden mountain is a merely possible entity.
The round square is an impossible entity.

(ALF-pred) is valid only if we consider the quasi-formalized form, but not the superficial grammatical form: only colloquial sentences of the same semi-formalized form should be formally symbolized by applying the same symbolization schema. More specifically, some sentences entail the commitment to the existence of the referents of its singular and/or object terms, others not. A further elaboration of this distinction between sentences with and sentences without existential force may help us solve the problem of formalization inadequacy.

4 Committing and non-committing sentences

Here a principle I am not willing to abdicate: Any adequate theory of formalization must be sensitive to the intended meaning of colloquial expressions. Thus, it may be helpful to consider the distinction between committing and non-committing sentences. Committing sentences entail existential force. For a typical committing sentence is a sentence whose truth is or would be grounded on the fact that the referent(s) of the (singular or general) subject term exists and is the way the predicate says it is (or is not); or, with other words, the referents instantiate the property expressed by the predicate. In order to instantiate the property, first of all, the referents must exist. A non-committing sentence is a sentence that does not entail existential force of (some of) its terms. Their truth or falsity is either grounded on the non-existence of some of its referents or, at least, compatible with its non-existence. This is not a definition, but a characterization. Here some examples of intuitively true and false committing sentences (I apologize for probably giving excessively many examples in this section, but as Wittgenstein (1953: d 593) correctly warned: 'a main cause of philosophical disease is an unbalanced diet: one nourishes one's thinking with only one kind of example'):

The present King of France is bald. (F)
The present Queen of England is bald. (F)
Donald Trump is American. (T)
Angela Merkel is American. (F)
Some monkey devours some raisins. (T)
Some monkey devours some craisins. (F)
Some monkey devours every craisin. (F)

> *No kmonkey devours some raisins. (T)*
> *Jane bought blue bananas at the market. (F)*
> *Monkeys are insects. (F)*
> *I saw craisins yesterday. (F)*

These are committing sentences, whose truth, in the case they are true, is grounded on the fact that the referred entities (in subject or object position) are the way the predicate state they are and whose falsity, in the case they are false, is grounded either (i) on the fact that the referred entities (in subject or object position) exist but are not the way the predicate describe them to be or (ii) because they simply do not exist.

There are basically two kinds of non-committing sentences: (i) explicitly non-existence sentences and (ii) existence independent sentences. Examples of (i) are:

> *Pegasus does not really exist. (T)*
> *Angela Merkel does not really exist. (F)*
> *The present King of France does not exist. (T)*
> *The present Queen of England does not exist. (F)*
> *Harry Potter is a merely fictional character. (T)*
> *The golden mountain is a merely possible entity. (T)*
> *There are no raisins. (F)*
> *There are no craisins. (T)*

These sentences explicitly state the non-existence of some putative entities. They are true when the referred entity does not exist, as explicitly stated, and false otherwise. Most of them are not connected with formalization difficulties. Finally, here are some examples of true existence independent sentences, which constitute the most interesting case:

> *Big Foot was never found by the expeditors. (T)*
> *I have never been invited for tea with the present King of France. (T)*
> *I have never been invited for tea with the present Queen of England. (T)*
> *I never saw kmonkeys. (T)*
> *I never saw monkeys. (F)*
> *I have never eaten raisins. (F)*
> *I have never been to China. (T)*
> *I have never been to the Hogwarts School. (T)*
> *I did never climb the Sugar Loaf. (F)*
> *Little Augusto is still waiting for Santa Claus. (T)*
> *Some people believe in any kind of gods. (T)*

The monster of lake Ness was never seen. (T)
God has never answered my prayers. (T)
Ponce de León was searching for the Fountain of Youth. (T)

Some of these sentences are true because the referents exist and are the way the sentence says they are, e.g. I exist and so does China and I have never been there. However, some of these sentences are arguably true not because their referents exist and are the way the sentences say they are, but because their referents simply do not exist. Big Foot was never found not because he is well-hidden, but simply because 'he' does not exist; I never saw kmonkeys not because they live far away from me, but because there are none; and poor little Augusto is not still waiting for Santa Claus because Santa Claus is late this years, but simply because he does not exist. Four our purpose here, the interesting features of these sentences is that they do not seem to carry any ontological commitment to the existence of a referent of some of its referential terms. Contrary to the committing sentences, the existence independent sentences do not imply the existence of all referents of the referential terms of the sentence for being possibly true.

Since the truth conditions of existence independent sentences are fairly ambiguous, their logical form is far more controversial. Unfortunately, existence independent sentences are not easily detectable as explicitly non-existence sentences. Indeed, I do not think to be possible to formulate a universal and purely syntactical rule for identifying existence independent sentences. Natural language is too capricious to permit such a clear cut-off distinction. In particular, it seems that there cannot be a purely syntactical test for distinguishing committing and non-committing sentences. In the examples above, the non-committing character depended not on syntactical but on semantical features, viz. the meaning of predicates like 'was never found', 'has never been seen', 'never been to' and 'has been waited for'. So, for solving the inadequacy problem, instead of focusing on the difference between instantiated and uninstantiated predicates like 'raisin' and 'craisin' we should focus on the committing force of the whole sentence, which includes the relational predicate 'devour'. Jacquette noted correctly that it sounds false to say that *some monkey devour every craisin* and I think this is due not to the use of the predicate 'craisin', but to the semantical properties of 'devour' in a sentence of the form 'some Fs devours every G'. The mere use of the predicate 'craisin' by itself has no existential force, as sentences like 'there are no craisins' and 'I wish there were craisins' trivially witness.

Probably, the best we can do for capturing the existential force of a sentence is to execute the careful step-by-step paraphrase procedure. Now, I also want to suggest here what I hope to be a useful semantical test for helping to detect the non-committing character of some sentences. We may recognize that a given predicate does not entail existential force appealing to the following Gricean-styled cancelation test: take the sentence and add a cancelation of the existence of the referent of the term at stake with a 'simply because it does not exist' or 'although it does not exist' clause (where the 'it' refers to the term at stake). In the case of existence independent non-committing sentences this yield a perfectly reasonable — although not necessarily true — sentence, since the existence is not really entailed but at most merely suggested. Let's apply the test to our non-committing sentences above:

> *Big Foot was never found by the expeditors simply because he does not exist. (T)*
> *I have never been invited for tea with the present King of France simply because he does not exist. (T)*
> *I have never been invited for tea with the present Queen of England simply because she does not exist. (F)*
> *I never saw kmonkeys simply because they do not exist. (T)*
> *I never saw monkeys simply because they do not exist. (F)*
> *I have never eaten raisins simply because they do not exist. (F)*
> *I have never been to China simply because this place does not exist. (F)*
> *I have never been to the Hogwarts School simply because this place does not exist. (T)*
> *I did never climb the Sugar Loaf simply because this place does not exist. (F)*
> *Little John is still waiting for Santa Claus simply because Santa Claus does not exist. (T)*
> *Some people believe in any kind of gods, although they do not exist. (T)*
> *The monster of lake Ness was never seen, simply because it does not exist. (T)*
> *God has never answered my prayers, simply because he does not exist. (who knows?)*
> *Leon is searching for the font of youth, although it does not exist. (T)*

The cancelation of the existential force of these sentences sounds not artificial or contradictory at all, although it is false in some cases. This may be seen as an indication for the fact that these sentences do not entail existential force. The opposite: the cancelation of existence is so natural that in some cases it is better introduced not by a 'although', but by the grounding clause 'because' or 'simply because'. This is so because the non-existence of the referents is not only compatible with the sentences, but in some cases (not in all) even a partial or full ground for its truth. The same test applied to committing sentences yields a quite different result:

> *The present King of France is bald simply because he does not exist. (?)*
> *The present Queen of England is bald simply because she does not exist. (?)*
> *Donald Trump is American simply because he does not exist. (?)*
> *Angela Merkel is American simply because she does not exist. (?)*
> *Some monkey devours some raisin simply because they (raisins) do not exist. (?)*
> *Some monkey devours some craisin simply because they (craisins) do not exist. (?)*
> *Some monkey devours every craisin simply because they (craisins) do not exist. (?)*
> *Jane bought blue bananas at the market simply because there are no blue bananas. (?)*
> *Monkeys are insects simply because there are no monkeys. (?)*
> *I saw craisins yesterday simply because there are no crasins. (?)*

Note that according to this test, not the general terms 'monkey', 'raisin' or 'craisins' carry the ontological commitment by themselves, but the predicate 'devour', 'is', 'are', 'was bought', etc. However, we should not conclude that, given a particular predicate F, F will have the same committing or non-committing character in all possible occurrences. Our test is sufficiently fine-grained to be able to capture this distinction. Compare e.g. the two sentences which are syntactically equivalent

> *Jane is still waiting for the perfect husband simply because he does not exist.*
> *Anne is still waiting for her husband simply because he does not exist.*

These sentences are quite instructive: 'waiting for x' (like 'searching for x', 'looking for x' and many other verbs) is in some cases, but not in others, not committing to the existence of x: Jane will plausibly wait forever for the perfect husband simply because no real man is good enough for her. Cancelation works perfectly well in this sentence. A similar cancelation of the Anne-sentence sounds fairly odd although it uses the same predicate in a sentence with a similar structure. The obvious explanation is: the difference between both sentences emerges because of the definite description 'her husband', which by itself carry the ontological commitment in this particular context.

To sum up: the ALF claims that colloquial expressions of the same logical form can and should be formally symbolized by applying the same symbolization schema. This is false at the superficial level of the grammatical form. At the deep level this may be true, but in order to know which is the 'real' logical form of a sentence, a simply syntactical analysis is not enough. We must consider contextual and semantical aspects of the sentence in order to decide its logical form, in particular its committing or non-committing character. Committing sentences, explicitly non-committing sentences and existence independent sentences have different logical forms due to some of their particular syntactical and semantical features.

5 Formalization of General Terms

Now we are finally in position to propose a solution to the problem of expressive inadequacy. In fact, the solution is simple: we must reject assumption (ii)

(ii) 'Some monkey devours every raisin' is correctly translated as $\exists x[Mx \land \forall y[Ry \to Dxy]]$;

In the ordinary situation, whenever we say that *devours(some F, every G)*, we are committing ourselves to the existence of Fs and Gs, for this is clearly a committing sentence. And this is due to the standard meaning of 'devour'. Let's apply our cancelation test to (S1) and (S2)

(S1)-test Some monkey devours every raisin simply because there are no raisins.

(S2)-test Some monkey devours every craisin simply because there are no craisins.

It seems clear that both sentences failed the non-existence cancelation test. It should strike any competent speaker as odd that some monkey devours every raisin or craisin simply because these do not exist.

Note that Jacquette himself in his presentation of the paradox revels a similar intuition, one concerning the existence of the referent of the subject term and one concerning the existence of the referent of the object term. Concerning the object of the sentence, he writes that, since there are no craisins, it sounds false to say that some monkey devours every craisin: 'none of the existent monkeys devours any let alone every craisin, since craisins there are none' (p. 25). Concerning the subject term, as we saw, he considers that

(T0) $\exists x \forall y[[Mx \land Ry] \to Dxy]$

is not an adequate formalization of

(S1)   Some monkey devours every raisin.

because (T0) is not committed to the existence of any monkeys, contrary to the colloquial (S1). Indeed, the cancelation test of (S1) fails blatantly: the sentence

Some monkey devours every raisin simply because/although there are no monkeys.

is clear non-sense. Therefore, according to Jacquette himself, a sentence of natural language with the structure $R(\text{some } F, \text{every } G)$ entails that there are both Fs and Gs. But then, we ask, why should we take in our formalization the existential commitment of the subject term seriously, but not that of the object term? 'Ontological correctness' requires that both must be taken in consideration in formalization. Therefore, in my view the sentence

(S1)   Some monkey devours every raisin.

should not be translated as

(T1) $\exists x[Mx \land \forall y[Ry \to Dxy]]$,

but as

(F1) $\exists x[Mx] \land \exists x[Rx] \land \exists x \forall y[[Mx \land Ry] \to Dxy]$.

I do really think that this is in accordance with common use of natural language: we usually do not say that *some* monkey devours *every* raisin unless we think that there are monkeys and raisins. Thus, the following paraphrase procedure seems adequate:

(S1)   Some monkey devours every raisin.

(S1)*   There are monkeys, and some of them devour every raisin.

(S1)** There are monkeys and raisins, and some of these monkeys devour every one of that raisins.

(F1) $\exists x[Mx] \land \exists x[Rx] \land \exists x \forall y[[Mx \land Ry] \to Dxy]$.

How does this new formalization strategy solve the expressive inadequacy problem? As we saw, the problem emerged because the intuitively false sentence

(S2)   Some monkey devours every craisin

is usually translated as

(T2) $\exists x[Mx \land \forall y[Cy \to Dxy]]$,

which is classically true, while the apparently true natural sentence

(S4) No monkey devours every craisin,

is traditionally translated as

(T4) $\neg\exists x[Mx \land \forall y[Cy \to Dxy]]$

which syntactically contradicts (T2) and therefore is false (pace Jacquette). And so, what should be true is false, and what should be false is true.

With our proposal for formalizing *devours[some φ, every ψ]* the formalization of the intuitively false sentence (S2) becomes

(F2) $\exists x[Mx] \land \exists x[Cx] \land \exists x \forall y[[Mx \land Cy] \to Dxy]$.

which is, as expected, false, since the condition '$\exists x[Cx]$' is not satisfied. Further, as expected, the syntactical contradiction of (F2), i.e.

($\neg$F2) $\neg[\exists x[Mx] \land \exists x[Cx] \land \exists x \forall y[Mx \land Cy] \to Dxy]]$

is true, since it is the negation of the false (F2). Indeed, it is not the true that some monkey devours every craisin. We might be tempted to consider ($\neg$F2) an adequate translation of (S4). But it is certainly more correct to consider that, since (F2) is a complex sentence, it admits several different internal negations. And it is no obvious that (S4) is intuitively true, as Jacquette suggests — at least not in any context. Suppose a competent speaker who is confronted with the assertion (S4). Under the supposition that he believes in what is said, he will probably conclude that there are monkeys and craisins and that no monkey is able to devour every craisin, i.e. he will interpret (S4) as

(F4) $\exists x[Mx] \land \exists x[Cx] \land \neg\exists x \forall y[[Mx \land Cy] \to Dxy]$,

which is false, since the second conjunct is false. But the same speaker, as soon as he get the information that in reality there are no craisins, will probably not consider (S4) true, but simply misleading. The sentence which we would consider true is either 'it is not true that some monkey devour every craisin', or 'no monkey devours every craisin simply because there are no craisins' or, even better 'no monkey devours any craisin simply because craisins do not exist at all'. Jacquette probably considers (S4) intuitively true because he has one of these variants in mind.

Anyway, both (F2) and (F4), like their colloquial counterparts, are false, since both entail the existential force ($\exists x[Cx]$). And contrary to standard formalization, in the framework of the new formalization strategy what should be true is true, and what should be false is false. The formalization inadequacy is overcome.

6 One Objection and three Advantages

Of course, my proposal may be challenged in many ways. I think its most serious weakness consists in the fact that the cancelation test is based on a quite vague notion of non-sense. It is not always clear whether the cancelation test yields a reasonable sentence or not. One may for example dispute whether my example of a non-sensical sentence

Anne is still waiting for her husband simply because he does not exist.

is really non-sensical. Most philosophers are quite ingenious for creating unexpected scenarios. One could e.g. imagine that poor Anne has a fantasy prone personality and just imagine having a husband despite the fact that she has never been married. This way, the sentence sounds not that odd. But I think that even in this case, the speaker who reports the situation would not use this misleading sentence, but something like 'poor Anne, she is still waiting for her imaginary husband!' or similar.

Further, and more important: the cancelation test is not intended as a universal and infallible procedure for specifying the exact meaning of any sentence independent of the context. On the opposite, my first goal here is just warning that, in order to figure out the logical form of a particular statement we must take into account the intended meaning of the speaker. Only the speaker himself may be able to decide if her statement e.g. of the form $R(some\ F, every\ G)$ is intended to be committing to Fs and Gs or not. So, the best advice I can give: ask the speaker! The step-by-step paraphrase procedure and the cancelation test were only proposed here as useful devices for deciding, in the absence of context and speaker, if a given sentence has existential force or not. Anyway, in some particular cases, the test may deliver quite plausible results, as the monkey-craisin sentence shows.

I also believe that my proposal has some advantages. First, it solves the inadequacy problem in the framework of classical logic without being ad hoc. I honestly think it helps capturing better the intended meaning of natural speakers. Logic textbooks very often try to impose upon us the artificial view that some clear instances of committing sentences are examples of non-committing or existence independent sentences. So any logic teacher has to struggle a lot in order to make digestible to beginners in logic that both sentences

All unicorns are black.
All unicorns are white.

are at the same time true. The discomfort of natural speakers confronted with this oddness does not dissipate when we explain that these sentences are true because the antecedent of the implication 'if something is a unicorn, than it is F' is false or because the set of unicorns is the empty set, which is contained in any set, including in the set of white things and in the set of black things. And, honestly, the rhetoric question 'did you ever saw a unicorn that is not black? No?! Do you see, there are no unicorns that are no black. Therefore, all unicorns are black!' (the same for white) does not really help much.

According to my suggestion, we do not have to change logical or set theoretical principles or go non-classic to overcome this difficulty. In fact, not classical logic is fault, but our usual formalization. All we have to do is to repeat the recipe of Russell's history of success with his theory of descriptions: we have to recognize the existential force of committing sentences like 'the F is G', 'some Fs devour every Gs' and 'all Fs are Gs'. Standardly, when we say that all Fs are Gs or that some Fs devour every Gs we assume the existence of Fs and of Gs. We do not have to change logic and its way, but the way we grasp the intended meaning of competent speakers when we translate their sentences into logic.

A second advantage consists is this: the proposed formalization offers a more fine-grained logical expression of our thoughts. For instance, it makes possible to distinguish different grounds for the falsity of a sentence. In our example,

(F2) $\exists x[Mx] \wedge \exists x[Cx] \wedge \exists x \forall y[[Mx \wedge Cy] \rightarrow Dxy]$.

may be false because of three different reasons: it may be false because (i) there are no monkeys ($\neg \exists x[Mx]$), (ii) because there are no craisins ($\neg \exists x[Cx]$) or, finally, (iii) because of the limited digestive capacity of monkeys, i.e. because of the third part of the conjunction ($\neg \exists x \forall y[Mx \wedge Cy] \rightarrow Dxy]$) (or any combination of these reasons, of course). The usual formalization (T2) only recognized possibilities (i) and (iii). Further, this more fine-grained structure of the formalization in (F2) perfectly correspond to our intuitions in natural language about which particular content entailed in a false sentence is the ground for its falsity. In our example,

(S1)   Some monkey devours every raisin.

is not false because there are no monkeys or because there are no raisins, but just because of the limited digestive capacity of monkeys. Contrary to this, the sentence (S2) is not false because of the limited digestive capacity of monkeys, but because there are no craisins , i.e. because '$\exists x[Cx]$' is false.

Here the third advantage: my proposal offers a possible solution to the old paradox of the square of opposition. According to the traditional view, the square of oppositions represents a collection of logical relationships between the four categorical forms

A— Every S is P
E— No S is P
I— Some S is P
O— Some S is not P

The paradox emerges when we suppose that 'S' is an empty term. In this case, the I form is false, for it states that 'some S is P'. Consequently, its contradictory E form 'no S is P' must be true. But then the subaltern O form 'some S is not P' must be also true — but it is classically false! For, how can O be true, if there are no Ss? (See for this Kneale and Kneale 1962: 58-9 and Parsons 2008)

The most obvious reaction consists in rejecting empty terms in syllogistic. This was defended for example by Lukasiewicz (1950), Strawson (1950: 343-4), Geach (1950: 480), Patzig (1968:6-7) and Smith (2009). Curiously, 'rejecting empty terms was never a mainstream option, even in the nineteenth century' (Parsons 2008). Note that our proposal for solving the expressive inadequacy problem does not consist in simply rejecting the use of empty terms. As we defended above, the assumption (iii) should be accepted: uninstantiated predicates can enter into logical formalism. Our proposal consists rather in explaining how to translate more accurately the intended natural meaning of colloquial sentences in general. And here there are two options.

Applying our test, we might conclude that the four categorical forms of syllogistic are committing sentences with intuitive existential force. It seems that

A— All Ss are Ps simply because/although there are no Ss.
E— No Ss are Ps simply because/although there are no Ss.
I— Some Ss are Ps simply because/although there are no Ss.
O— Some Ss are non-Ps simply because/although there are no Ss.

are non-sense. Therefore, we could propose the corresponding formalizations:

A— $\exists x[Sx] \wedge \exists x[Px] \wedge \forall x[Sx \to Px]$
E— $\exists x[Sx] \wedge \exists x[Px] \wedge \forall x[Sx \to \neg Px]$
I— $\exists x[Sx \wedge Px]$
O— $\exists x[Sx \wedge \neg Px]$

As a result, if there are no Fs, the I form becomes false. Of course, now the E form is no longer the contradictory form of I, neither is the A form the contradictory of O. And since E entails the existence of S it is false in the case S is empty, just as the O form as well. If we want to keep the square with the traditional logical relations, we could propose the following formalizations

A— $\exists x[Sx] \wedge \exists x[Px] \wedge \forall x[Sx \to Px]$
E— $\exists x[Sx] \wedge \exists x[Px] \wedge \forall x[Sx \to \neg Px]$

I— $\neg[\exists x[Sx] \wedge \exists x[Px] \wedge \forall x[Sx \to \neg Px]]$
O— $\neg[\exists x[Sx] \wedge \exists x[Px] \wedge \forall x[Sx \to Px]]$

However, as a second and more reasonable option, we may accept that, contrary to A, E has no existential force, since the cancelation of A sounds much more nonsensical than the cancelation of E. If we additionally accept that for Aristotle the O form has no existential force, since his more usual formulation was 'not every S is P' instead of 'some S is not P', we get a formalization that corresponds to Read's (2015: 5) proposal, according to which

A— $\exists x[Sx] \wedge \forall x[Sx \to Px]$
E— $\neg\exists x[Sx \wedge Px]$
I— $\exists x[Sx \wedge Px]$
O— $\exists x[Sx \wedge \neg Px] \vee \neg\exists x[Sx]$

This way, uninstantiated predicates are allowed in syllogistic and a consistent square of opposition with all its usual logical relations is preserved.

Conclusion

In this paper, classical predicate logic was defended. I argued that a more close analysis of Jacquette's paradox reveals that it is not a genuine paradox. However, it reveals a deep inadequacy of standard methods of formalization of colloquial sentences into classical logic. We oriented our investigation on three assumptions of our formalization praxis. From this, we concluded that:

1. uninstantiated predicates can legitimately enter into (meaningful, true or false) predicates-quantificational symbolizations;

2. it should not make any difference in formalization if a predicate is instantiated or not, i.e. given two predicates F and G, one of them instantiated and the other not, any sentence of the equivalent grammatical form ('G' just substitutes any occurrence of 'F' in the sentence) must be formalized according to the same pattern. In this context, I distinguished existential committing sentences from existential non-committing and existential independent sentences and argued that this difference must be made explicit in formalization. Given the absence of a universal syntactical test for detecting the committing force of a colloquial sentence, I proposed a semantical cancelation test. In any case, it is not truth that any colloquial expressions of the same syntactical form can and should be formally symbolized by applying the same symbolization schema: it

depends on some more general semantical and contextual features. From this, I concluded that

3. contrary to usual opinion, the sentence 'some monkeys devours every raisin' is not correctly translated as $\exists x[Mx \wedge \forall y[Ry \rightarrow Dxy]]$, but as $\exists x[Mx] \wedge \exists x[Rx] \wedge \exists x \forall y[[Mx \wedge Ry] \rightarrow Dxy]$, for it clearly has existential force.

This way we avoid the necessity of going non-classic. Classical logic and the way it is is not inadequate. The expressive inadequacy problem lies rather in some vicious habits of our standard methods of formalization.

References

[1] Chateaubriand, O. (2001) Logical Forms. Part I. Truth and Description. Campinas: Centro de Lógica, Epistemologia e História da Ciência/ UNICAMP, 2001. (Coleção CLE, 34)

[2] Chateaubriand, O. (2002) 'Descriptions: Frege and Russell Combined'. Synthese 130: 213-226.

[3] Donnellan, K. (1966). 'Reference and Definite Descriptions'. The Philosophical Review vol. 75 (3): 281Ð304.

[4] Frege, G. (1982) 'Über Sinn und Bedeutung', in Zeitschrift für Philosophie und philosophische Kritik, 100: 25Ð50. Translated as 'On Sense and Reference' by M. Black in Translations from the Philosophical Writings of Gottlob Frege, P. Geach and M. Black (eds. and trans.), Oxford: Blackwell, third edition, 1980.

[5] Geach, P. (1950) 'Subject and predicate'. Mind 59:461-482.

[6] Jacquette, D. (2010) Logic and How it Get That Way, Acumen.

[7] Lukasiewicz, J. (1951). Aristotle's Syllogistic. Clarendon Press, Oxford.

[8] Lukasiewicz, J. (1957). Aristotle's Syllogistic. Clarendon Press, Oxford, 2nd enlarged edition.

[9] Parsons, T. (2008). 'The traditional square of opposition'. In Zalta, E. N., editor, The Stanford Encyclopedia of Philosophy. Fall 2008 edition.

[10] Patzig, G. (1968). Aristotle's Theory of the Syllogism. Reidel, Dordrecht.

[11] Read, Stephen (2015) 'Aristotle and Lukasiewicz on Existential Import'. Journal of the American Philosophical Association 1 (3): 535-544.

[12] Russell, B. (2005) 'On Denoting'. Mind 14: 479Ð493; repr. in Bertrand Russell, Essays in Analysis, London: Allen and Unwin, 1973, 103Ð119; and in Bertrand Russell, Logic and Knowledge, London: George Allen and Unwin, 1956, 41Ð56; also appearing in Collected Papers, Vol. 4.

[13] Smith, R. (2009). 'Aristotle's logic'. In Zalta, E. N., editor, The Stanford Encyclopedia of Philosophy. Spring 2009 edition.

[14] Strawson, P. (1950). 'On referring'. Mind, 59:320-344.

[15] Wittgenstein, L. (1953) Philosophical Investigations. G.E.M. Anscombe and R. Rhees (eds.), G.E.M. Anscombe (trans.), Oxford: Blackwell.

[16] Zalta, E. N. (1983) Abstract Objects. An Introduction to Axiomatic Metaphysics, Dordrecht: Reidel.

[17] Zalta, E. N. (1988) Intensional Logic and the Metaphysics of Intentionality, Cambridge: MIT Press.

How do we Know Things with Signs? A Model of Semiotic Intentionality

Manuel Gustavo Isaac
Swiss National Science Foundation, and University of Amsterdam, The Netherlands.
isaac.manuelgustavo@gmail.com

Abstract

Intentionality may be dealt with in two different ways: either ontologically, as an ordinary relation to some extraordinary objects, or epistemologically, as an extraordinary relation to some ordinary objects. This paper endorses the epistemological view in order to provide a model of semiotic intentionality defined as the meaning-and-cognizing process that constitutes to power of the mind to be about something on the basis of a semiotic system. After a short introduction that presents the components of semiotic intentionality (viz. sign, act, content, referent) along with their division into an intending and a fulfilling side (section 1), the first main part of the paper analyzes semiotic intentionality at its primary level (a.k.a. 'concrete intentionality') as a real and subjective relation between meaning-intending and meaning-fulfilling acts grounded in the manipulation of some semiotic system (section 2). Then, building on such concrete intentionality, the second main part of the paper analyzes semiotic intentionality at its secondary level (a.k.a. 'abstract intentionality') as an ideal and objective relation between intentional and fulfillment contents, which in turn: (i) proceed from an abstraction performed on the intending and fulfilling acts, respectively, and (ii) retroactively categorize the intending and fulfilling acts, respectively (section 3). Finally, from this combination of an act-based conception of content with a presentationalist account of intentionality, the conclusion of this paper produces the

---

This research paper is published within the project 'Sign, Meaning, Reference: A Semiotic Epistemology for Conceptual Engineering' funded by the Swiss National Science Foundation (Project Number 2SKP1_171736). Versions and parts of this paper have been presented at the Discourse in Philosophy Colloquium and the Amsterdam Metaphysics Seminar (University of Amsterdam), 'The World in Us' conference (University of Edinburgh), and the 9th European Congress of Analytic Philosophy (University of Munich), and the Cambridge Mind Seminar (University of Cambridge). I am grateful to those who provided comments on these occasions. Special thanks to the Amsterdammers Franz Berto, Jelle Bruineberg, Ilaria Canavotto, Peter Hawke, Julian Kiverstein, Martin Lipman, and Tom Schoonen.

intended model of semiotic intentionality in such a way that knowledge and truth are then respectively defined in it as the subjective correspondence between the two acts of concrete intentionality and as the objective correspondence between the two contents of abstract intentionality (section 4).

Keywords: Phenomenology, Intentionality, Semiotic epistemology, Act-based contentfulness, Presentationalism.

Prolegomena

Intentionality may be dealt with in two different ways: either ontologically, as an ordinary relation to extraordinary objects (e.g. as a naturalizable relation to socalled immanent or inexistent objects, be they termed 'mental representations,' 'intentional contents,' 'Fregean sense,' etc.), or epistemologically, as an extraordinary relation to ordinary objects (e.g. as a non-causal relation to some transcendent object, whatever their ontological status may be).[1] The ontological take corresponds to the mainstream approach to intentionality in current analytic philosophy of mind. It originates in the Brentanian famous reintroduction of the notion[2] and is strongly connected with a representationalist understanding of the mind's aboutness (which is usually assumed as the unquestioned common ground of all standard debates around intentionality).[3] Against the current, this paper endorses the so-called epistemological view on intentionality—which, from now on, will be defined as the meaning-and-cognizing process that constitutes the power of the mind to be about something. Focusing on the mode of meaning-and-cognition that makes use of some semiotic system to direct the mind toward something as significant (namely, on semiotic intentionality), it aims at providing a systematic overall framework for the study and interpretation of such semiotic referential meaning

---

[1] On the conceptual distinction between these two types of intentionality (without their labelling, though), see [9, Chap. 6 [113]]—whereas by "ontological" and "epistemological", I mean here ontologically and epistemologically committing, respectively.

[2] For the record: "Every mental phenomenon is characterized by what the Scholastics of the Middle Ages called the intentional (or mental) inexistence of an object, and what we might call, though not wholly unambiguously, reference to a content, direction toward an object (which is not to be understood here as meaning a thing), or immanent objectivity. Every mental phenomenon includes something as object within itself, although they do not do so in the same way. [...] This intentional in-existence is characteristic exclusively of mental phenomena. No physical phenomenon exhibits anything like it. We can, therefore, define mental phenomena by saying that they are those phenomena which contain an object intentionally within themselves" [5, 68].

[3] Another source of this banal view lies in the West Coast reading of Husserl [23].

process considered as a cognitive process (in other words, a model of semiotic intentionality). The bold question that the paper addresses in this regard is then: How do we mean-and-cognize things with signs? And this question has in turn two correlated issues, namely: first, that of the accessibility of the semiotic semantics (i.e. what is the subject matter of the semiotic intentionality and how do we access it?); second, that of the reliability of the semiotic cognition (i.e. what is the quality criterion of semiotic intentionality and how do we secure it?). The strategy adopted to tackle these issues, and thereby achieve the main aim of this paper, consists in a phenomenological analysis of the structure and processes of semiotic intentionality taken as the meaning-and-cognizing relationship that links a signifier to something that is signified, determining in this way the mind's aboutness.[4] With this in mind, the paper starts with a short introduction in which the components of semiotic intentionality (viz. sign, act, content, referent) are presented, along with their division into an intending and a fulfilling side (section 1). Next, the first main part of the paper analyzes semiotic intentionality at its primary level (a.k.a. 'concrete intentionality') as a real and subjective relation between meaning-intending and meaning-fulfilling acts grounded in the manipulation of some semiotic system (section 2). Then, building on such concrete intentionality, the second main part of the paper analyzes semiotic intentionality at its secondary level (a.k.a. 'abstract intentionality') as an ideal and objective relation between intentional and fulfillment contents (section 3). And finally, this paper concludes with the intended model of semiotic intentionality designed in such a way that: (i) the intentional and fulfillment contents (viz. the subject matter of semiotic intentionality) are accessed in it as the byproducts of an abstraction performed on the intending and fulfilling acts, respectively, and (ii) knowledge and truth (viz. the quality criteria of semiotic intentionality) are then respectively defined in it as the subjective correspondence between the two acts of concrete intentionality and as the objective correspondence between the two contents of abstract intentionality (section 4). So in the end, such a model of semiotic intentionality will amount to the combined defense of an act-based conception of content along with a presentationalist account of intentionality.[5] And what will be thereby uncovered is what structures the power of the mind to be about something.

---

[4]*Nota bene.* This analysis builds on the theory of intentionality developed by Husserl in the *Logical Investigations*; yet, it is in no way an interpretative reconstruction of Husserl (neither systematic, nor historical).

[5]In this vein, see [2, 3, 6, 10, 12, 16, 24, 29].

1 Introduction

Semiotic intentionality starts with a simple binary semiotic model in which a signifier is linked to something that is signified via some sort of relationship. Building on this simple binary model, it then proceeds to its recasting in the light of intentionality defined as the meaning-and-cognizing process that constitutes the power of the mind to be about something. On those grounds, the signifier/signified pair is relabeled 'signifying/signified' (to put an emphasis on the meaning process that is already at work with the object-source that is operating as a signifier [subsection 2.1]),[6] whereas its internal relationship is analyzed into two components, namely the act and the content. As for these two components, they are in turn subdivided into two phases: first, that of the intention proper (i.e. the source-to-target relationship); second, that of the corresponding fulfillment (i.e. the target-to-source relationship). So that in this way, the structural articulation in play within this intentionalist semiotic model can be characterized as an act linking a source to a target (and backward) via some interaction with a content. A semiotic model built in this framework thus comprises four structural components (viz. sign, act, content, and referent), along with four relational ones (viz. the intending and fulfilling acts, plus the intentional and fulfillment contents). This paper deals with unpacking what makes the core of such model, that is: the intentional relationship (see Table 1).

	SIGNIFYING	SIGNIFIED	
SOURCE	\multicolumn{2}{c}{INTENTIONAL RELATIONSHIP}	TARGET
Sign	\multicolumn{2}{c}{Act}	Referent
	Intending act *First phase →*	Fulfilling act *← Second phase*	
	Intentional content	Fulfillment content	
	\multicolumn{2}{c}{Content}	

Table 1. The components of semiotic intentionality

---

[6]Thanks to Alistair Isaac for having pointed this out.

2 Concrete intentionality

This section forms the first main part of the paper and it deals with concrete intentionality, that is with intentionality taken as a real relation. This relation links two actual components, namely: the intending acts and the fulfilling acts (subsection 2.2, 2.3). And these two actual components are themselves considered as constituting the subjective intimation of the meaning-and-cognizing process that intentionality is. In the case of semiotic intentionality, such subjective intimation is grounded in a semiotizing process, which is itself at the principle of any proper use of a semiotic system as such—that is, of any meaning-and-cognizing use of a semiotic system as an artificial device (subsection 2.1).[7] While displaying in this way the primary/lower level of semiotic intentionality, this section will provide the concrete subjective basis necessary and sufficient for producing, at a secondary/upper level, the abstract objective content of intentionality (section 3).

2.1 Semiotizing process

In the case of semiotic intentionality, semiotization is the trigger for intentionality as a meaning-and-cognizing process. Such operation starts with a remote object taken as a concrete real phenomenon, and it processes this object in such a way that this object then becomes able to serve as signifier. The operation of semiotization is nothing but this constitution of an object as a signifier.[8] And this constitution itself consists in an intentional shift: one and the same thing is successively endowed with two different meaning-and-cognitive status and function. First, when it is the mere object that is about to become a signifier, it is intuitively given as the target of an intention (viz. the something that the mind is about when the mind directs itself toward this thing as significant) (see Figure 1a). Then, in a second step, when the same thing operates as a signifier, it is no longer the target, but instead the source of an intention (itself oriented toward or aiming at some further intended

---

[7]Given such pragmatic anchoring in the world of cognitive devices that the semiotic systems are, semiotic intentionality is intrinsically connected to the 4E approaches to cognition according to which: cognitive systems/processes are grounded in, or even partly constituted by, bodily activities (*Embodied cognition*), which are themselves located in external environments (*Embedded cognition*) that are in turn involved in cognition (*Extended cognition*) through their being-brought-forth-as-significant by the cognitive agents (*Enacted cognition*)—more specifically, semiotic intentionality is compatible with the so-called strong embodied and embedded theses as promoted by the cognitive integration framework [20, 21, 22].

[8]The canonical example here is that of an arabesque being deciphered, and therefore becoming apt to work and be used as a signifier (viz. a meaningful sign)—this can be extended, though, to anything being recognized as being potentially meaningful (e.g. a mark in the ground, a MD prescription, a smell in the air, etc.).

target) (see Figure 1c). In the phenomenological tradition at the background of the current analysis of semiotic intentionality, the operator that performs this intentional shift has been dubbed 'intentional consciousness.' Intentional consciousness is in this sense what supplements the immediate awareness of something (in the present case, the thing that is about to operate as a signifier) by an interpretative operation — that is, an operation that takes this something as some other thing that it is not immediately (see Figure 1b).[9] This taking-as operation involves a special cognitive apprehension, which turns out to correspond to the arbitrary activation of the potential meaning-and-cognizing function of something — while the arbitrariness of such activation means that no ontological constraints limits what can serve as a sign in the first place (the only further limitation here will be semantic, determining as such how a sign can operate as signifier [subsubsection 3.3.1]). Thereby, through such intentional semiotization process, something is constituted as a semiotic tool by and for the intentional consciousness (see Figure 1). And when this tool is then put into use, it runs in two phases, the intending one first (subsection 2.2), and next the corresponding fulfilling one (subsection 2.3).

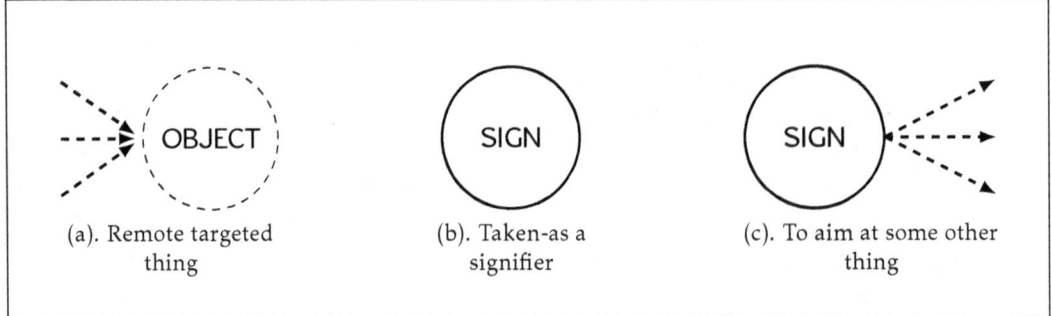

Figure 1. First stage of semiotic intentionality[10]

2.2 Intending process

The intending phase of the process of semiotic intentionality concerns the source-to-target relationship. It only consists in the intending acts that give a meaning-

---

[9]It may worth noting here that intentional consciousness is also nothing more than this: a functional operator that performs a phenomenal transformation (viz. a transformation in the mode of appearance of something).

[10]Note here that the act does not come out of the sign per se, but only as an object that has been processed by the intentional consciousness; in other words, a sign always is something taken-as some other thing (namely, a signifier) by the intentional consciousness (in short: sign = something + consciousness).

and-cognizing function to an object by using it as a signifier to aim at some referent (see Figure 2).[11] These intending acts as such are firstly characterized by directedness, that is, by an exclusive direction toward their intended referent. In this regard, they operate as access providers to their intended referent as targeted in and by them. And consequently, they merge in themselves the meaning-of and the meaning-on something (viz. the meaning they express with what this meaning is about—or, to put it another way, the intentional and the intended object). Secondly, these intending acts are characterized by a form of relationalism. Such relationalism means here that these acts are nothing else than a linking process that orientate a source on its target as provided in and by the very same acts themselves. The status of these acts is in this respect that of a pure transitivity: while they carry meaning-and-cognition, they are the meaning-and-cognition, and no medium obstructs them as carriers of intentionality, providers of referential relationship. As a consequence of this second characteristic of the intending acts, what is here strongly rejected is any ternary model of the semiotic referential meaning process in which a sign is articulated to its referent via a medium (a.k.a. 'mental representations,' 'intentional contents,' 'Fregean sense,' etc.) that then constitutes an intermediary step from the sign to its referent. Rather, in the intentionalist model of semiotic referential meaning, meaning-and-cognition starts by existing only in acts. Now the theory of intentionality faces here is a critical tension between, on one hand, a relation (namely, the intending act) that prevails over its relata that it determines in their role and status (i.e. the being source and target of the object-sign and the referent, respectively), and on the other hand, the relata on which depends the relation itself (i.e. the being a relation of the intending act). Intentionality, indeed, is such extraordinary relationship. The rest of the paper is dedicated to make sense of its original tension (cf. section 4).

2.3 Fulfilling process

The fulfilling phase of the process of semiotic intentionality concerns the target-to-source relationship. It consists in the acts that fulfill the intention of an intending-act with the intended referent. To understand when and where these acts take part in the semiotic referential meaning-and-cognition process, two senses of intentionality have to be distinguished and defined, each of which corresponds to

---

[11] As a follow-up to the previous example (see note 8), the example here is that of the deciphered arabesque being used and working as a signifier to aim at its referent (e.g. Allah in the case of an arabesque of the Throne verse, etc.)—this can be extended of course to every meaning-and-cognitive uses of any semiotic system working as such via the involvement of one of our five senses (e.g. a speech, a pictogram, a flavor, etc.).

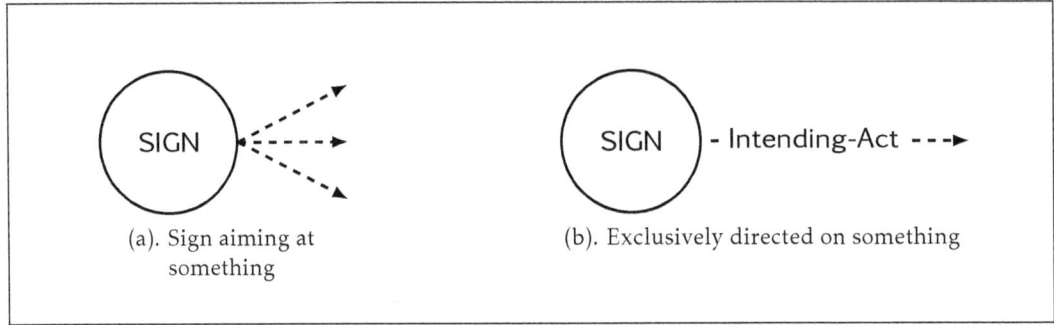

Figure 2. Second stage of semiotic intentionality

a different possible outcome the semiotic referential meaning-and-cognition process, constituting as such two successive steps of this very same process. So, first is the narrower sense of intentionality. In this case, the intention is the empty intention of an unfulfilled intending-act. The target referent is merely intended, without being given in intuition, neither sensorily nor non-sensorily. As a target, its accessibility is therefore not even properly or effectively realized. And intentionality is in this respect (still) an irrelative concept: that of a relation with only one relatum (see Figure 3a).[12] Building upon those grounds, the second sense of intentionality is a wider one. In this case, the intention is the complete intention of a filled intending-act. The target referent is no longer merely intended, but rather also given in intuition, sensory or not. As a target, its accessibility is therefore properly and effectively realized. And intentionality is in this respect (now) a relative concept: that of a relation with two relata, correlating thereby the act of aiming at something with that of hitting the mark as intended (see Figure 3b).[13] In the framework of these two definitions, the fulfilling phase of the process of semiotic intentionality clearly belongs to the wider sense of intentionality. And as such, the intuitive givenness, sensory as well as non-sensory, of the intended target referent is bound to remain supererogatory in the process of semiotic referential meaning-and-cognition. It will always only take part in this process as an illustration, a confirmation, a reinforcement of the primary source-to-target direction in

---

[12]To follow-up again on the example of the newly meaningful arabesque (see note 11), this would concern an intrinsically unreachable referent (e.g. Allah presumably in the case of an arabesque of the Throne verse, etc.) — more spontaneous examples may include any uses of proper names (or sentences) without intuitive givenness (sensory or non-sensory) of their supposedly correlated object (or state-of-affairs, respectively).

[13]Here, spontaneous examples would be that of any uses of proper names (or sentences) with intuitive givenness, sensory or non-sensory, of their correlated object (state-of-affairs, respectively) (cf. note 12).

the whole structure of intentionality, which will always remain first and foremost an aboutness relationship.

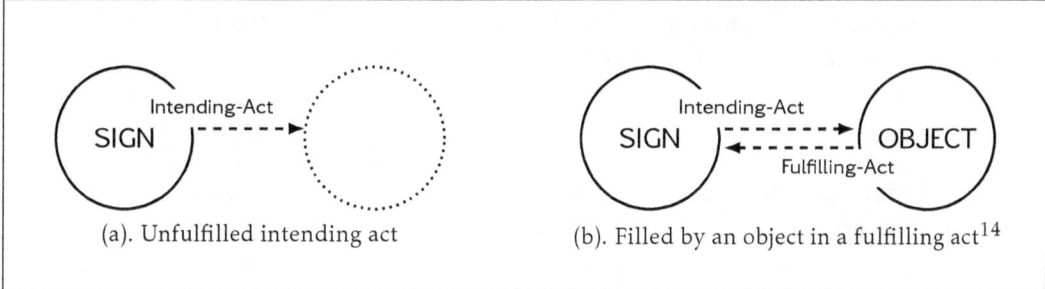

Figure 3. Third stage of semiotic intentionality

3 Abstract intentionality

This section forms the second main part of the paper, which deals with abstract intentionality, that is with intentionality taken as an ideal relation (cf. section 2). This relation links two contentual components, namely: the intentional contents and the fulfillment contents. And these two components are themselves considered as constituting the abstract objective correlates of the meaning-and-cognizing process that intentionality is (ibid.). As such, they are based on acts and obtained as by-products through a threefold process of abstraction (see Remark 3.1). Now, while analyzing the three steps of this threefold abstraction process — from generalization (subsection 3.1) to formalization (subsection 3.3) via idealization (subsection 3.2) — that constitutes the secondary/upper level of intentionality, this section will show how this very process retroactively categorizes the primary/lower level of intentionality, thereby applying the strategy of a regressive transcendental argument (see Proposition 3.2).[15]

Remark 3.1 (Abstraction). *In each of its three modes (viz. generalization, idealization, formalization), the semiotic intentional abstraction consists in the act of seizing upon a*

---

[14]This is an explicit way in which the present model of semiotic intentionality diametrically differs from the Husserlian one in which the fulfilling acts come out of some mental agent (source-to-target direction as well, then), without any mediation of the sign-object in itself (cf. note 4): here, the fulfillment is rather to be understood in receiver terms — I am grateful to one of the reviewers for having drawn my attention to this distinctive feature.

[15]On the terminological distinction between the so-called regressive transcendental argument and its opposite, the progressive one, see [1].

common property of a group of things — or in other words, in grasping a commonality among them, which is then possibly taken as an identical objective unit;[16] *through its sequencing in three steps, it progressively constitutes the contents of intentionality as resulting from a rationalization process that makes progressively explicit the inferential articulations of such contents of intentionality.*

3.1 Generalization

Generalization, or generalizing abstraction, is the first of the three steps of the abstraction process that constitutes the contents of intentionality in the form of a rationalization progress (see Remark 3.1). The inputs of this first mode of abstraction generally comprise first-order objects. In the case of semiotic intentionality and with respect to the constitution of the intentional and fulfillment contents, these first-order objects correspond to the intending and fulfilling acts, respectively. The operation of generalizing abstraction itself then consists in focusing on similarities to be extracted. And the outputs of such extraction process will be called here abstract 'species' or 'genera.' On the *source-to-target direction* of the intentionalist semiotic model, these so-called species are intentional species that consist of the structural invariant of some indefinite plurality of intending acts (subsection 2.2): their status is that of an equivalence relation on these intending acts [EQREL(1)],[17] while their function is to stabilize the source-to-target relationship itself (see Figure 4a).[18] Whereas, on the *target-to-source direction* of the intentionalist semiotic model, these species are fulfillment species that consist of the structural invariant of some indefinite plurality of fulfilling acts (subsection 2.3): their status is that of an equivalence relation on these very same fulfilling acts [EQREL(2)],[19] while their function is to stabilize symmetrically the target-to-source relationship (see Figure 4b [also note 18]). The secondariness of these two types of species therefore relies in their characterization as by-products of a reflection on concrete intention-

---

[16] Hence, abstracta are always by-products, and their status is thus that of secondariness: they are non-independent contents (a.k.a. 'parts') distinguished from other such contents of a concretum (a.k.a. 'whole').

[17] Consequently, such intentional species are radically immanent in the sense that they do only exist in act(s).

[18] The stabilizing function of generalizing abstraction is the reason why going for the abstract and contentful secondary/upper level of intentionality instead of remaining with its concrete and contentless primary/lower level—thanks to Richard Menary for having raised this issue.

[19] Consequently, such fulfillment species, which only occur in the case of complete intention of filled intending-acts, are not radically transcendent in the sense that they always only represent the target referent hit-as-intended in the first place by the intending acts.

ality.[20] And as such by-products, they provide the abstract objective basis for the thematization of the two sides of intentionality as ideal unities (subsection 3.2).

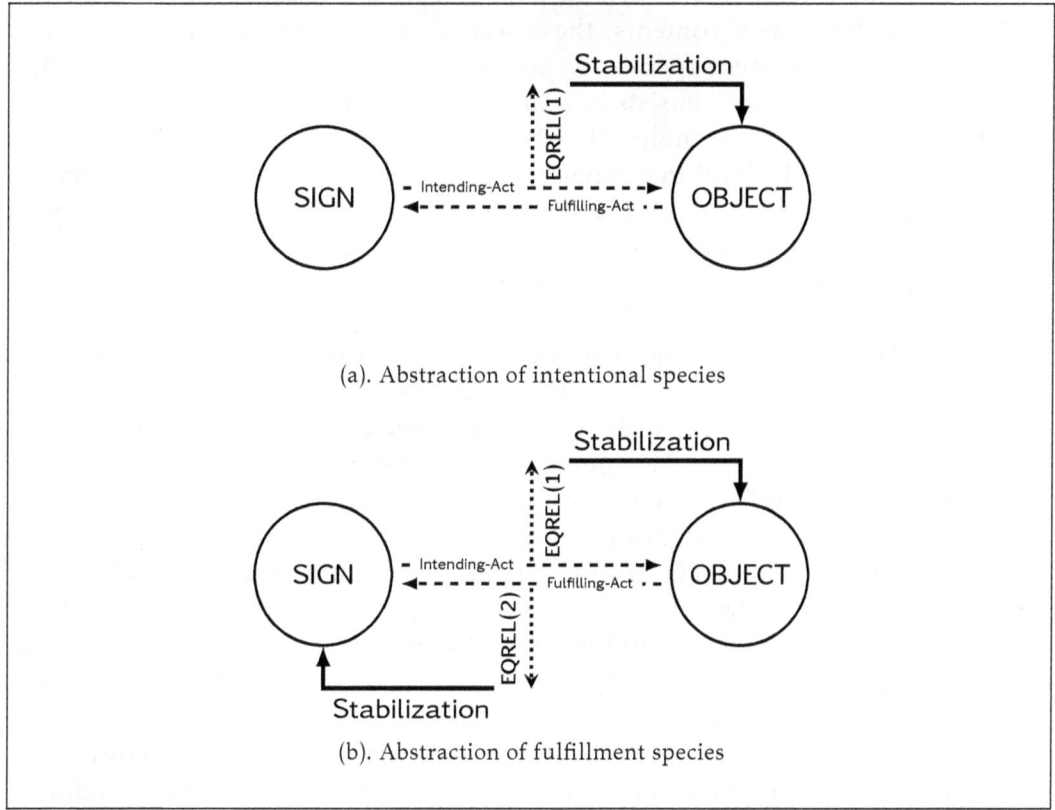

Figure 4. Fourth and fifth stages of semiotic intentionality

3.2 Idealization

Idealization, or idealizing abstraction, is the second of the three steps of the abstraction process that constitutes the contents of intentionality in the form of a rationalization progress (see Remark 3.1). The inputs of this second mode of abstrac-

---

[20]For examples of intentional species, one may think of concepts defined either in terms of exemplarist theories of concepts according to which conceptual structures are bodies of information about some particular exemplary instance of a concept to which resemble the other members of its extension (e.g. Fido for the concept 'dog,' a Ford Mustang for the concept 'car,' etc.) or in terms of prototypist theories of concepts according to which such structures are lists of typical properties statistically/probabilistically satisfied by the members of the concept's extension (e.g. quadruped, hairy etc. for the concept 'dog,' four-wheeled, mobile etc. for the concept 'car,' etc.).

tion comprise the second-order objects that are the species or genera that have been produced at the first step of the intentional abstraction process (subsection 3.1). In the case of semiotic intentionality and with respect to the constitution of the intentional and fulfillment contents, these second-order objects correspond to the intentional and fulfillment species, respectively (ibid.). The operation of idealizing abstraction itself then consists in a thematization of the equivalence relations that the act species are (see notes 21, 22)—which as such amounts to an asymptotic progress toward a limit that is neither realized, nor realizable by any member of the group of things on which the abstraction is performed. As for the outputs of such thematization process, they will be called here abstract 'idealities.' On the *source-to-target direction* of the intentionalist semiotic model, these idealities are intentional idealities that consist of unified intentional species (cf. Figure 4a): their status is that of equivalence classes of intending acts resulting from the partition of some plurality of intending acts (viz. the quotient set of the set of intending acts) by the equivalence relation that the intentional species are [EQCLAS(1)],[21] while their function is precisely to partition (that is, to categorize) the set of intending acts (see Figure 5a). Whereas, on the *target-to-source direction* of the intentionalist semiotic model, the so-called idealities are fulfillment idealities that consist of unified fulfillment species (cf. Figure 4b): their status is that of equivalence classes of fulfilling acts resulting from the partition of some plurality of fulfilling acts (viz. the quotient set of the set of fulfilling acts) by the equivalence relation that the fulfillment species are [EQCLAS(2)],[22] while their function is precisely to partition (that is, to categorize) the set of fulfilling acts (see Figure 5b). The secondariness of these two types of idealities therefore relies in their characterization as by-products of a second-order reflection on concrete intentionality.[23] And as such by-products, they provide the abstract objective basis for the formalization of the two sides of intentionality as categorial unities (subsection 3.3).

---

[21] Taken to be here a thematizing operation, the use of an equivalence relation to partition a set of things produces its outputs, the equivalence classes (a.k.a. 'idealities'), as intransitive conceptual objects.

[22] As in the case of the source-to-target direction (see note 21), such equivalence classes are output as intransitive objects—yet, this time, as intransitive phenomenal objects (which are in this respect the ideal correlates of the intended and fulfilling referents [subsection 2.3]).

[23] For examples of intentional idealities, one may think of concepts now defined in terms of definitionist theories of concepts according to which conceptual structures are definitional complexes that are logically composed of more basic semantic features, which in turn provide the necessary and sufficient conditions for an object to fall under them (e.g. the concepts of 'dog' and 'car' as found in a dictionary) (cf. note 20).

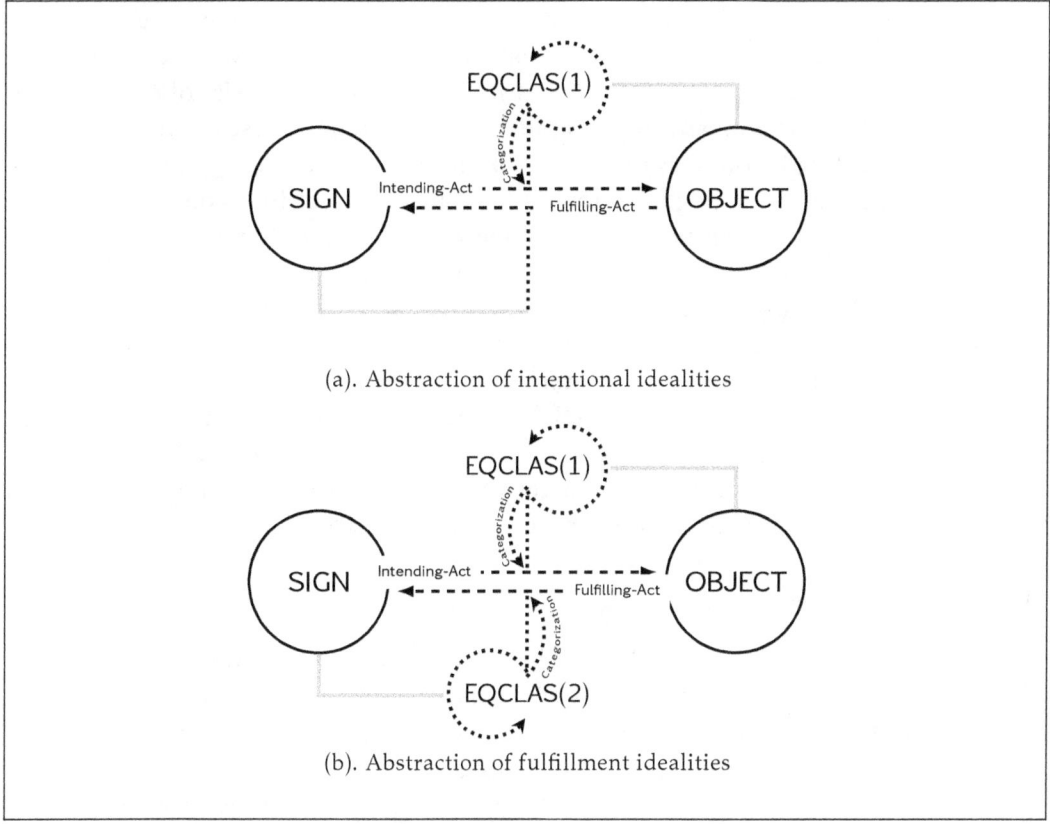

Figure 5. Sixth and seventh stages of semiotic intentionality

3.3 Formalization

Formalization, or more precisely formalizing abstraction, is the third and last of the three steps of the abstraction process that constitutes the contents of intentionality in the form of a rationalization progress (see Remark 3.1). The inputs of this third mode of abstraction generally consist of higher-order objects that are either the species/genera or the idealities, which have themselves been produced at the first and second steps of the intentional abstraction process, respectively (subsection 3.1, 3.2). In the case of semiotic intentionality and with respect to the constitution of the intentional and fulfillment contents, these first and second-order objects on which the formalizing abstraction is performed correspond to the intentional and fulfillment species or idealities, respectively (ibid.). The operation of formalizing abstraction itself then consists in isolating the formal features, either of the equivalence relations that the act species are, or of the equivalence classes that the act idealities are (in other words, in removing their content to grasp their pure form).[24] As for the outputs of such isolation process, they will be called here abstract 'categorialities' (a.k.a. 'categories'). Categorialities consist of empty categorial forms of intentional species and idealities. And when these empty categorial forms are considered from a systemic viewpoint, the network they integrate forms a categorial domain.[25] Categorial domains are networks of interconnected empty categorial forms whose connections are nomologically regulated by law-like determinations of any possible categorial combination (viz. any combination of categorialities).[26] Now, as in the two first steps of the intentional abstraction process (cf. subsection 3.1, 3.2), such outputs come in two types, that of the intentional categorialities and that of the fulfillment ones along with their respective categorial domains (subsubsection 3.3.1, 3.3.2), matching thereby the two sides of the intentionalist semiotic model. And the secondariness of these two types of categorialities relies in turn in their characterization as by-products of a higher-order (viz. second or third) reflection on concrete intentionality.[27]

3.3.1 Transcendental apophantics

On the source-to-target direction of the intentionalist semiotic model, the categorialities are intentional categorialities that consist of empty categorial forms of intentional species and idealities (see Figures 4a, 5a), and such categorialities belong to what will be called here the 'apophantic' categorial domain (see Figure 6a). In this so-called apophantic domain, the law-like determinations for the possible combinations of intentional categorialities are layered on two levels. On a *lower level*, the

---

[24] This use of the notion of 'form' is in fact the very reason why calling this third and last mode of abstraction the 'formalizing' one (or in short, 'formalization').

[25] Note that it is also possible to consider both the species/genera and the idealities that are obtained in the two first steps of the semiotic intentional abstraction process with respect to the networks they integrate; but in the systematic overall framework of semiotic intentionality, the status and function of such networks won't commensurate with the transcendental operativeness of the categorial domains (subsubsection 3.3.1, 3.3.1)—which is what ultimately matters in view of the main aim of this paper (cf. Prolegomena).

[26] As made of categorial forms, categorial domains are themselves formal, and as governed by law-like principles that are independent from any content (especially, from any material content and any existential status), they are as well analytic—furthermore, with respect to the rooting of the threefold intentional abstraction process in a parthood relationship (see Remark 3.1), the structure of such domains is expected to be mereological.

[27] For examples of basic intentional categorialities, one may think of concepts now defined in terms of theory theories of concepts according to which conceptual structures are structured bodies of theoretical explanatory principles (causal, functional, generic, and nomological) about the members of the concept's extension (cf. notes 20, 23).

compatibilities under consideration are the morphological ones, and what is there determined is the possibility to obtain unitary (as opposed to non-unitary) combinations of intentional categorialities. The status of the law-like determinations in play is therefore that of formation rules. In the case of the apophantic domain, such rules operate a distinction between the sense and the nonsense.[28] And in this way, the function of these law-like determinations is to condition the source-to-target meaningfulness, that is: the very possibility for a source to aim at a target (viz. in the semiotic case, for a sign to vehicle a meaning) (cf. subsection 2.2 [see also: subsection 2.1]). Next, on an *upper level*, the categorial compatibilities of the apophantic domain that are under consideration are the syntactic ones, and what is determined on this level is the possibility to obtain consistent (as opposed to an inconsistent) combinations of intentional categorialities. The status of the law-like determinations that are now in play is therefore that of transformation rules. In the case of the apophantic domain, such rules subdivide the sense (singled out on the lower level) by distinguishing within it the consistent and the inconsistent sense (a.k.a. counter-sense).[29] And in this way, the function of these law-like determinations is to condition the source-to-target 'fillability', that is: the very possibility for a meaningful source to hit its target as intended (viz. in the semiotic case, for a meaningful sign to access its referent) (cf. subsection 2.2, 2.3). Via such two-layered nomological regulation of the apophantic categorial domain, the formalizing abstraction uncovers the immanent structure that transcendentally conditions [TRANSCON], and in a retroactive way, the intending acts of semiotic intentionality at its primary/lower level (see Figure 6b and Proposition 3.2)—given that this transcendental conditioning concerns, in the apophantic case, the source-to-target direction of the intentionalist semiotic model (a.k.a. the narrow sense of intentionality [subsection 2.3]), it is considered to be a direct conditioning of the overall framework of semiotic intentionality.

3.3.2 Transcendental ontology

On the target-to-source direction of the intentionalist semiotic model, the categorialities are fulfillment categorialities that consist of empty categorial forms of fulfillment species and idealities (see Figures 4b, 5b), and such categorialities belong to what will be called here the 'ontological' categorial domain (see Figure 7a). In this so-called ontological domain, the law-like determinations for the possible

---

[28] To take linguistic examples from both natural and formal languages, what is thereby ruled out is semiotic strings such as 'red is or,' '$+_x = a)(f,$' etc.

[29] To take once again linguistic examples from both natural and formal languages, what is ruled out here is meaning-carrying semiotic strings such as 'non-squared square,' '$p \land \neg p,$' etc.

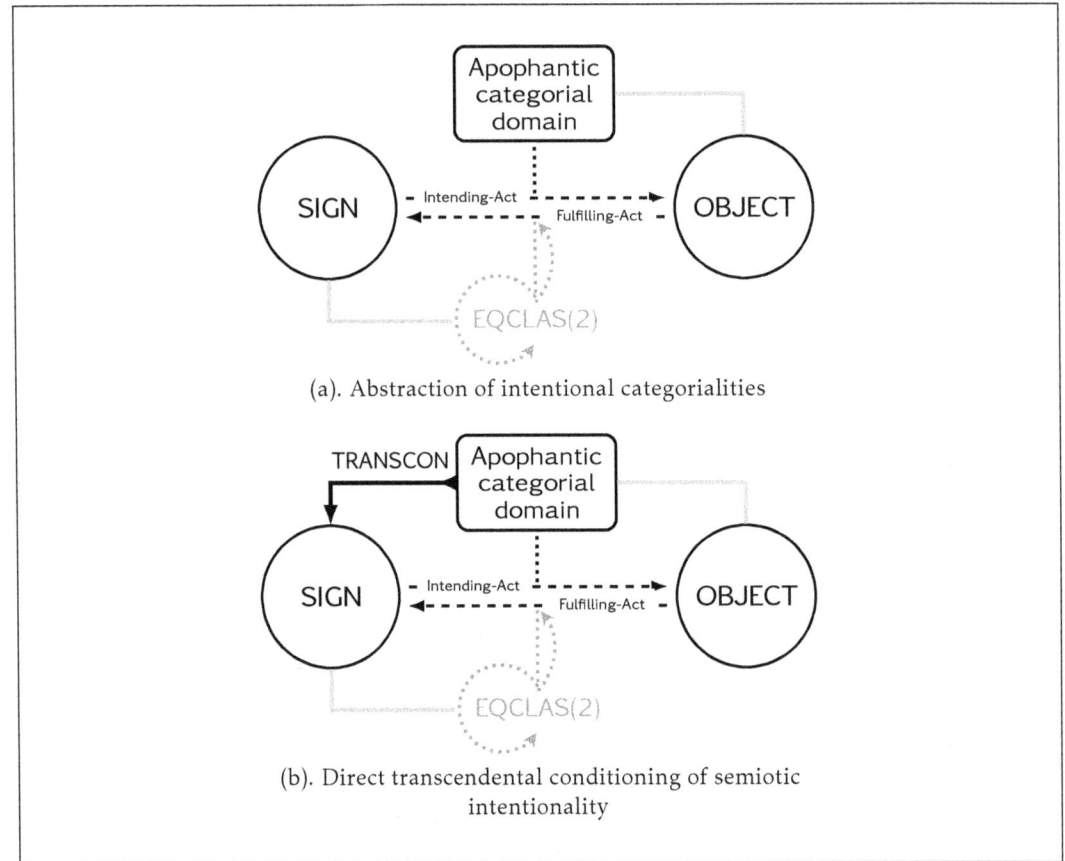

Figure 6. Eighth stage of semiotic intentionality

combinations of fulfillment categorialities are also layered on two levels. On a *lower level*, the compatibilities under consideration are again the morphological ones, and what is determined there as well is the possibility to obtain unitary (as opposed to non-unitary) combinations of fulfillment categorialities. So, as in the apophantic case, the status of the law-like determinations in play here is that of formation rules. Yet in the ontological domain, such rules operate a distinction between the targetable and the untargetable.[30] And in this way, the function of these law-like determinations is to condition the target-to-source targetability, that is: the very possibility for something to be the target of an intending act (cf. sub-

---

[30]What is thereby ruled out is arguably such composites as the monstrous mereological fusions entailed by the so-called 'unrestricted composition principle' [17, 18] (e.g. when a substance is attributed to an attribute that is itself taken as a substance).

section 2.2, 2.3). Next, on an *upper level*, the categorial compatibilities of the ontological domain that are under consideration are again the syntactic ones, and what is determined on this level is the possibility to obtain consistent (as opposed to an inconsistent) combinations of fulfillment categorialities. As well as in the apophantic case, the status of the law-like determinations that are now in play is therefore that of transformation rules. Yet in the ontological domain, such rules subdivide the targetable (singled out on the lower level) by distinguishing within it the consistent and the inconsistent targetable things—namely, between the intuitively givable (sensorily or not) on one hand, and the intuitively ungivable (sensorily or not) on the other hand.[31] And in this way, the function of these law-like determinations is to condition the target-to-source 'fillingness', that is: the very possibility for some targetable thing to be fulfilling (i.e. intuitively given, sensorily or not) (cf. subsection 2.3). Via such two-layered nomological regulation of the ontological categorial domain, the formalizing abstraction uncovers the immanent structure that transcendentally conditions [TRANSCON], and in a retroactive way, the fulfilling acts of semiotic intentionality at its primary/lower level (see Figure 7b and Proposition 3.2)—given that this transcendental conditioning concerns, in the ontological case, the target-to-source direction of the intentionalist semiotic model (a.k.a. the wider sense of intentionality [subsection 2.3]), it is considered to be an indirect conditioning of the overall framework of semiotic intentionality.

Proposition 3.2 (Transcendental regressive argument). *In the systematized overall framework of semiotic intentionality, the primary/lower level of semiotic meaning-and-cognition (viz. concrete intentionality) is retroactively conditioned, and in transcendental terms, by the secondary/upper one it produces (viz. abstract intentionality)—or, to put things the other way around, the secondary/upper level retroactively conditions, and in transcendental terms, the primary/lower one that produces it.*

4 Conclusion

This paper has endorsed an epistemological viewpoint on intentionality to focus on the mode of meaning-and-cognition that makes use of some semiotic system to direct the mind toward something as significant. Within this remit, its main aim was to provide a systematic overall framework for the study and interpretation of such semiotic referential meaning process considered as a cognitive process. The intended model of semiotic intentionality has been achieved through a phenomeno-

---

[31]What is ruled out here is arguably some targetable things such as the contradictory state-of-affairs that contravene the so-called 'ontological formulation' of Aristotle principle of contradiction [19] (i.e. the being-such-and-not-such of something under fully identical circumstances).

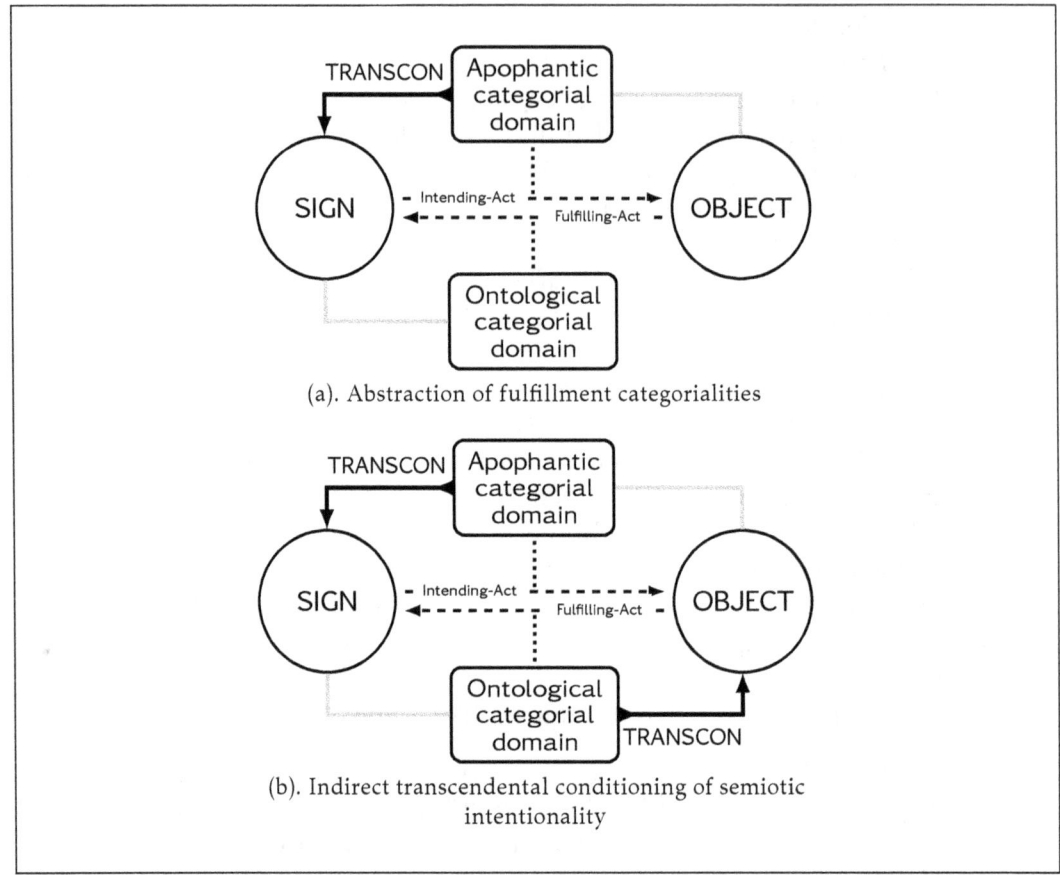

(a). Abstraction of fulfillment categorialities

(b). Indirect transcendental conditioning of semiotic intentionality

Figure 7. Ninth stage of semiotic intentionality

logical analysis of the meaning-and-cognitive structure and processes of semiotic intentionality. And the result takes the shape of the combination of an act-based theory of content with a presentationalist account of intentionality. According to this combination: (i) the contents of intentionality (intentional and fulfillment contents, respectively) are produced by a threefold abstraction performed on the acts of intentionality (intending and fulfilling acts, respectively);[32] whereas (ii) the acts of intentionality are themselves retroactively determined by the resulting contents of intentionality via the stabilization (subsection 3.1), the categorization (subsection 3.2), and the transcendental conditioning (subsection 3.3) of the semiotic ref-

---

[32]An ancillary upshot of such act-based theory of content lies in a simple solution to the highly topical and acute problem of the continuity from contentless to contentful meaning-and-cognition in the philosophy of (4E) cognitive science (e.g. [13, 14].

erential relationship (see Figure 8).[33] Against this background, the answer to the question of 'How do we mean-and-cognize things with signs?' addressed by this paper is pretty straightforward: we mean-and-cognize things with signs by the construction of the semantic artifacts (viz. the contents of intentionality) that enable us to shape and edit reality in order to make it intelligible — while the meaning-and-cognition process that intentionality is can now be conceived in presentationalist terms as an information modeling process [8]. In this framework, the subject matter of semiotic meaning-and-cognition then consists of nothing but the semantic artifacts that model our world-views — such subject matter being itself accessed as it is constituted, that is: through an abstraction process (cf. the accessibility issue of the semiotic semantics [Prolegomena]). As for the quality criterion of semiotic meaning-and-cognition, it consists of two original correspondences, that is: on a concrete subjective level, between the intending and fulfilling acts of intentionality (which may be termed 'knowledge'), and on an abstract objective level, between the intentional and fulfillment contents of intentionality (which may be termed 'truth')[34] — such correspondences being themselves secured by the informational optimization of the meaning-and-cognition process that intentionality is (cf. the reliability issue of the semiotic cognition [ibid.]). And on those grounds, the model of the semiotic mode of meaning-and-cognition (viz. the model of semiotic intentionality) may be generalized as a model of any mode of meaning-and-cognition (i.e. a semiotic model of intentionality). While determining the very conditions of possibility of any intentional meaning-and-fulfillment, such model indeed means nothing else than to uncover what structures the power of the mind to be about something.[35] Yet, as they say:

> If you have built castles in the air, your work need not be lost; that is

---

[33] An ancillary upshot of such presentationalism lies in the rejection of any kind of representationalism defined as involving both the dichotomy of the representing and the represented and the prevalence of the latter over the former (cf. the "strong sense" of re-presentation defined in [28, 135]) — another motive to reject representationalism here is that intentional abstracta are (inter-)personal level entities whose constitution derives from/depends on the worldly engagement of the cognitive agent (cf. note 7 and [25]'s typologies of anti-representationalisms).

[34] The originality of these two correspondences consists in their orthogonality to the standard *adaequatio rei et intellectus* — viz. in a verticality opposed to the old-fashioned horizontal *adaequatio* (see Figure 8).

[35] Ready-made implementations of the model of semiotic intentionality may thus contribute to: an account of the subject matter of formalisms [7], of the genesis of abstract objects [30], of the relationship of conceivability and possibility [4], of the distinction between aboutness and topicality [11], between so-called conceptual and phenomenal contents, a new step toward an integrative hybrid theory of concepts (see notes 20, 23, 27) [15], a renewed take on the symbol grounding problem (cf. [26, 27], etc.

where they should be. Now put the foundations under them. (Henry David Thoreau)

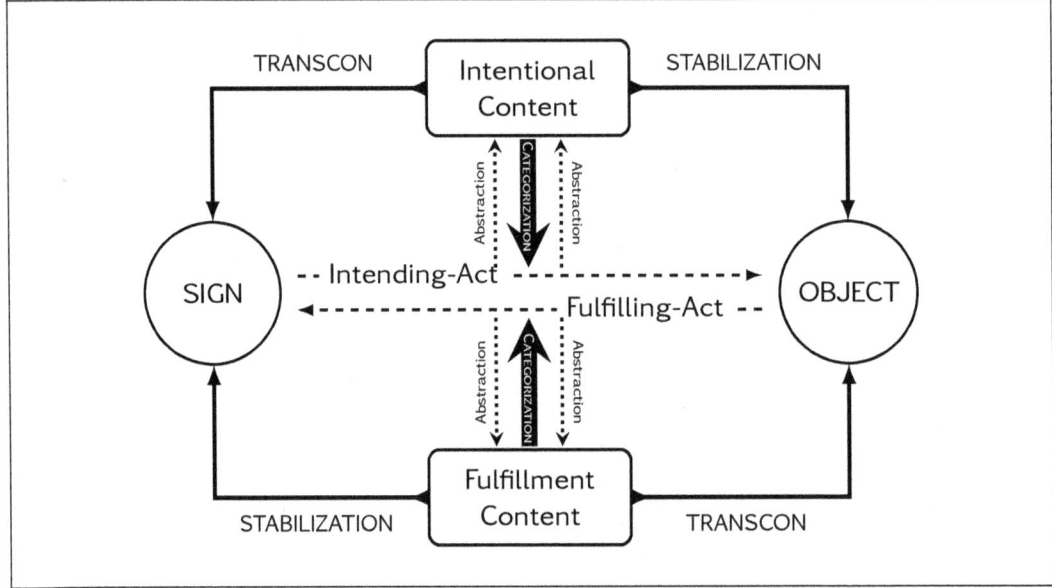

Figure 8. The intentionalist semiotic model

References

[1] Karl Ameriks. Kant's transcendental deduction as a regressive argument. *Kant-Studien*, 69(1-4):273–287, 1978.

[2] Jocelyn Benoist. *Phénoménologie, sémantique, ontologie. Husserl et la tradition logique autrichienne*. Presses universitaires de France, Paris, 1997.

[3] Rudolf Bernet. Husserl's Begriff des Noema. In Samuel IJsseling, editor, *Husserl-Forschung und Husserl-Ausgabe*, pages 61–80. Kluwer, Dordrecht, 1990.

[4] Francesco Berto and Tom Schoonen. Conceivability and possibility: Some dilemmas for Humeans. *Synthese*, pages 1–19, 2017.

[5] Franz Brentano. *Psychology from an Empirical Standpoint*. Routledge, London, 1973.

[6] John J. Drummond. Intentionality without representationalism, pages 115–133. Oxford University Press, Oxford, 2012.

[7] Catarina Dutilh Novaes. *Formal Languages in Logic*. Cambridge University Press, Cambridge, 2012.

[8] Luciano Floridi. A defence of constructionism: Philosophy as conceptual engineering. *Metaphilosophy*, 42(3):282–304, 2011.

[9] Shaun Gallagher and Dan Zahavi. *The Phenomenological Mind*. Routledge, London, 2008.

[10] Aron Gurwitsch. Husserl's theory of intentionality in historical perspective, pages 24–57. The Johns Hopkins University Press, Baltimore, 1967.

[11] Peter Hawke. Theories of aboutness. *Australasian Journal of Philosophy*, pages 1–27, 2017.

[12] Richard H. Holmes. An explication of Husserl's theory of the noema. *Research in Phenomenology*, 5:143–153, 1975.

[13] Daniel D. Hutto and Erik Myin. *Radicalizing Enactivism: Basic Minds without Content*. MIT Press, Cambridge (MA), 2013.

[14] Daniel D. Hutto and Erik Myin. *Evolving Enactivism: Basic Minds Meet Content*. MIT Press, Cambridge (MA), 2017.

[15] Frank C. Keil. *Concepts, Kinds, and Cognitive Development*. MIT Press, Cambridge (MA), 1989.

[16] Lenore Langsdorf. The noema as intentional entity: A critique of Føllesdal. *The Review of Metaphysics*, (37):757–784, 1984.

[17] David Kellogg Lewis. *On the Plurality of Worlds*. Blackwell, Oxford, 1986.

[18] David Kellogg Lewis. *Parts of Classes*. Blackwell, Oxford, 1991.

[19] Jan Łukasiewicz. On the principle of contradiction in Aristotle. *The Review of Metaphysics*, (24):485–509, 1971.

[20] Richard Menary. *Cognitive Integration: Mind and Cognition Unbounded*. Palgrave Macmillan, Basingstoke, 2007.

[21] Richard Menary. Intentionality, cognitive integration and the continuity thesis. *Topoi*, 28(1):31–43, 2009.

[22] Richard Menary. Mathematical cognition: A case of enculturation. In Thomas K. Metzinger and Jennifer M. Windt, editors, *Open MIND*, chapter 25(T). MIND Group, Frankfurt, 2015.

[23] David Woodruff Smith and Ronald McIntyre. *Husserl and Intentionality: A Study of Mind, Meaning, and Language*. Kluwer, Dordrecht, 1982.

[24] Robert Sokolowski. Intentional analysis and the noema. *Dialectica*, (38):113–129, 1984.

[25] Pierre Steiner. Enacting anti-representationalism. The scope and the limits of enactive critiques of representationalism. *Avant*, 5(2):43–86, 2014.

[26] Mariarosaria Taddeo and Luciano Floridi. Solving the symbol grounding problem: a critical review of fifteen years of research. *Journal of Experimental & Theoretical Artificial Intelligence*, 17(4):419–445, 2005.

[27] Mariarosaria Taddeo and Luciano Floridi. A praxical solution of the symbol grounding problem. *Minds and Machines*, 17(4):369–389, 2007.

[28] Francisco J. Varela, Evan Thompson, and Eleanor Rosch. *The Embodied Mind: Cognitive Science and Human Experience*. MIT Press, Cambridge (MA), 1991.

[29] Dan Zahavi. *Husserl's Phenomenology*. Stanford University Press, Stanford, 2003.
[30] Edward N. Zalta. *Abstract Objects: An Introduction to Axiomatic Metaphysics*. Springer, Dordrecht, 2012.

Schopenhauer's Consoling View of Death

Christopher Janaway
University of Southampton, UK.
C.Janaway@soton.ac.uk

1 Introduction

Dale Jacquette's open-mindedness and versatility as a philosopher are reflected in his extensive work on the philosophy of Arthur Schopenhauer.[1] I hope it will not seem inappropriate to begin from the contribution Jacquette chose to make to a volume on Schopenhauer that I edited some years ago, a piece entitled "Schopenhauer on Death".

Schopenhauer is pre-occupied with death to an extent that perhaps few modern Western philosophers have been. As Jacquette says, "Schopenhauer dwells morbidly on the inevitability of death" (1999: 298) and reminds us that every moment is a step nearer to it, even going so far as to call it the purpose or aim (*Zweck*) of life. Indeed, says Schopenhauer, living really is just a slow process of dying anyway:

> just as we know that walking is a continuously checked falling, the life of our body is only a constantly checked dying, a constant postponement of death ... Every breath we take wards off the perpetual onslaught of death; in this way, we struggle against death at every moment, and again at greater intervals, with every meal, every sleep, every time we warm ourselves, etc. Death has to win in the end: because we have been cast into death ever since birth, and it is only playing with its prey for a while before devouring it. (*WWR* 1, 337–8).[2]

The fear of death tends to be overwhelming for human beings, but Schopenhauer states that this fear cannot be rational (or at least wholly rational) in origin, for one thing because animals similarly fear death in the absence of any knowledge

---

[1] See items listed under "Jaquette" in the Bibliography.
[2] I use the following abbreviations for Schopenhauer's works: *BM*: *Prize Essay on the Basis of Morals*; *PP 2*: *Parerga and Paralipomena*, vol. 2; *WWR 1*, *WWR 2*: *The World as Will and Representation*, vols. 1 and 2.. For translations used, see Bibliography.

that they will die. The elemental fear of death must lie in something common to humans and other animals — this he calls the will to life (*Wille zum Leben*), arguing that it constitutes the essence of all living things. The will to life is not fundamentally a conscious *desire* for life. It is more akin to an innate disposition or drive, an elemental, pre-rational attachment or striving for life — and it is because we are an embodiment of this drive that the fear of life's ending is so ingrained in us. Nevertheless, Schopenhauer believes that this natural attachment to life need not have the last word, and offers us what he considers a consoling view of death. One considerable difficulty for the interpreter of Schopenhauer, however, is that the nature of the consolation on offer can be hard to fathom. On the one hand, Schopenhauer seems to be urging that death can be welcomed as an annihilation that releases us from the trouble involved in existing. On the other hand, he claims that something about us is not destroyed when we die. In what follows I shall seek to unpack this apparent duality by examining Schopenhauer's complex conception of the self.

Culturally prevalent views on this topic tend to "vacillate between the view of death as an absolute annihilation and the assumption that we are immortal, flesh, blood and all", says Schopenhauer, but, he continues, "Both are equally false: and we do not need to find a correct middle ground but rather a higher perspective from which both ideas fall away of their own accord" (*WWR* 2, 481). In Schopenhauer's view, the worst doctrine concerning death is that traditionally associated with Christianity, that each individual comes into existence "out of nothing" at some point in time, but then goes on existing forever. Against this he states:

> anyone who considers a person's birth to be his absolute beginning must consider his death to be his absolute end. For both are what they are in the same sense: consequently, we can only think of ourselves as immortal to the extent that we think of ourselves as unborn, and in the same sense . . . The assumption that a human being is created out of nothing leads necessarily to the idea that death is his absolute end. (*WWR* 2, 503–4)

Far preferable to Schopenhauer are the ideas he finds in Indian thought (though giving them Greek names): metempsychosis (transmigration of souls) and palingenesis (rebirth). He does not regard these doctrines as literally true, since in his ontology there are no such things as souls. However, as allegories, these doctrines graphically represent the idea that something fundamental to me both survives and predates the existence of the particular individual I identify myself as. But before we attempt to illuminate the nature of this surviving "something", let us address the central question whether, given Schopenhauer's views, we even *should* desire anything of ourselves to survive death.

2 Why not make life briefer?

Schopenhauer is famous for his claim that life consists of suffering, and indeed for his claim that because of this it would have been better not to have existed (see, e.g. *WWR* 2, 507–8; 588–92). This latter claim has often been treated, since the time of his greatest influence in the nineteenth century,[3] as the epitome of the pessimism with which his name is linked. And yet in his well-known treatment of suicide, Schopenhauer insists that it is a mistake to terminate one's life because of its sufferings. Taking his departure from this treatment of suicide, Jacquette wrote of "deep contradictions" in Schopenhauer's attitude to death:

> If the goal of philosophy is to reconcile individual will to the misery of existence and the elimination of consciousness in unreal death as the end and purpose of unreal life, then it appears impossible to explain why anyone should prefer death to a life of even the most acute turmoil, suffering, and pain. If the aim of life is death, and if death is unreal, then why should the philosopher not hasten to it? (Jacquette 1999: 313)

Leaving aside for the moment the notion that death is "unreal", Jacquette is here pointing to a kind of dilemma for Schopenhauer. On the one hand, it seems that death is something to be welcomed. It is, as Schopenhauer says, a "liberation" from life, and more dramatically "the great opportunity not to be I any longer" (*WWR* 2, 524). Hence one should say to a dying person, "you are ceasing to be something that it would have been better for you never to have become." (*WWR* 2, 517). But if so, why not take that opportunity as soon as you can? As Jacquette puts it, "once we get the picture, why not make life briefer?" (Jacquette 2000: 46). Assuming that suicide results in non-existence, it would seem inconsistent on Schopenhauer's part to claim that life is worse than non-existence, and to condemn suicide; yet this is what he does. For while he urges a sympathetic attitude towards those who take their own lives,[4] he nonetheless argues that intentionally bringing about one's own death is "morally" objectionable in one particular way:

> [T]he only relevant moral reason against suicide ... lies in the fact that suicide is counter to achieving the highest moral goal insofar as it substitutes a merely illusory redemption from this world of misery for the real one. But it is still a very long way from this error to a crime, which is what the Christian clergy want to make it out to be. (*PP* 2, 279)

---

[3] See Plümacher 1888: 124, for the view that philosophical pessimism as inaugurated by Schopenhauer is the view that non-being would be preferable to being.

[4] On the suicide of Schopenhauer's own father and its lasting influence on the philosopher, see Cartwright 2010: 87–94.

Jacquette conveys our likely response:

> Is Schopenhauer's conclusion necessitated by or even logically consistent with the concept of death he has elaborated? Or is Schopenhauer, having offered a powerful motivation for self-destruction, merely trying awkwardly now within his pessimistic and arguably nihilistic philosophical system to accommodate the squeamishness of traditional morality about the problem of suicide? (Jacquette 2000: 46)

However, to comprehend Schopenhauer's objection to suicide we need to probe further into his system of values. He speaks of suicide's hindering a "real redemption" from the world, and we must seek to understand what that is supposed to comprise. For Schopenhauer, it comprises what he calls the negation of the will to life. In a passage that has puzzled many readers, he explains that

> [f]ar from being a negation of the will, [suicide] is a phenomenon of a strong affirmation of will. This is because negation is not essentially an abhorrence of the *suffering* of life, but an abhorrence of its *pleasures*. The person who commits suicide wills life, and is only unsatisfied with the conditions under which life has been given to him. Thus, when he destroys the individual appearance he is relinquishing only life, not the will to life. He wills life, wills the unimpeded existence and affirmation of his body, but the tangle of circumstances does not allow him this and he undergoes great suffering (*WWR* 1, 425–6)

The crucial point to take from this is that there is more than one axis of evaluation in play for Schopenhauer. One issue is whether the individual continues existing or not. On this score, let us assume (provisionally at least) that for Schopenhauer — because of his negative evaluation of existence — there is positive value (for the individual) in the individual's ceasing to exist. But the other axis of value lies in whether one adopts the attitude of affirmation or negation towards life. Schopenhauer places the *highest* value on negation of the will to life, an attitude that we might also call *non-attachment* to the goods of ordinary individual existence. The human being, as a living thing, is naturally set up to have desires, wishes, impulses that tend towards the well-being or happiness of the bodily individual that they ordinarily identify themselves as. Schopenhauer captures the unreconstructed human outlook in a famous simile:

> Just as a captain sits in a boat, trusting the weak little vessel as the raging, boundless sea raises up and casts down howling cliffs of waves; so the human individual sits calmly in a world full of sorrow, supported by and

trusting in the *principium individuationis*, which is how the individual cognizes things as appearance (*WWR* 1, 379).

The *principium individuationis* comprises simply space and time: the individuality of a human being, is, like that of any other empirical thing, its distinctness in terms of spatio-temporal location. But as individual living things, and more specifically as individual human beings, we are egoistic by nature, in Schopenhauer's view. Egoism is endemic in every individual, as a central hard-wired disposition. It is "linked ... with his innermost core and essence, and indeed is properly identical with it," and in consequence "the human being unconditionally wills to preserve his existence, wills it unconditionally free from pains, ... wills every pleasure of which he is capable" (*BM*, 190). This fundamental orientation of human desire stems from the will to life that is our essence. Affirmation of the will to life is our default state, and is manifest simply in our pursuing our naturally arising desires. Non-human animals also affirm the will to life in this immediate way. But a more reflective affirmation of the will to life arises when human beings come to regard the attainment of desires centred on the well-being, or *happiness*, of the individual as the locus of genuine value.

But Schopenhauer argues that it is a mistake to locate genuine value there. Such happiness through individual desire-satisfaction is always receding from us and can never be complete. By contrast, will-lessness (*Willenslosigkeit*), or absence of will, has a finality about it that makes it of higher value, in Schopenhauer's view:

> we might figuratively[5] call the complete self-abolition and negation of the will, the true absence of will, the only thing that can staunch and appease the impulses of the will forever, the only thing that can give everlasting contentment, the only thing that can redeem the world, ... we might call this the absolute good, the *summum bonum*. We can look upon it as the one radical cure for the disease against which all other goods — such as fulfilled wishes and achieved happiness — are only palliatives, only anodynes. (*WWR* 1, 389)

For Schopenhauer the attainment of such a state of mind, which he here calls contentment (*Zufriedenheit*) and elsewhere blissfulness (*Säligkeit*),[6] is of the utmost importance. The relevance of this notion to our discussion of suicide is that, while Schopenhauer may hold that death is a desirable liberation for any human being, he

---

[5] I shall not discuss here the question why Schopenhauer thinks there can be a highest good only "figuratively". On this, see Janaway 2016; and, for alternative interpretations, Migotti 1995; Reginster 2012: 355–6.

[6] See e.g. *WWR* 1, 389, 401, 417, 419.

does not hold that all such cases of liberation are equally valuable, independently of their psychological context. Bringing about one's own death because one's ego-centred wishes are unfulfilled and one's happiness is unachieved will be a symptom of one's remaining attached to the will to life, of one's values remaining centred upon the individual and its desires. A person in such a state fails to attain the contentment (we might even say the enlightenment) that constitutes the highest good. Thus while extinction in death might be a good for the tormented individual, blissful detachment from his or her individual desires would be a higher good: and to attain such contentment one must be alive and face the suffering life brings.

In Schopenhauer's view, then, suicide attains a good, but if undertaken by the unenlightened, naturally egoistic person who has fallen into despair (and Schopenhauer makes the plausible implicit assumption that this is the most common kind of case), it cuts short the path to something more valuable. "Someone who commits suicide is like a sick person who, having started undergoing a painful operation that could cure him completely, does not allow it to be completed and would rather stay sick" (*WWR* 1, 427). The "painful operation" that can be a step towards negation of the will life is a life of suffering, and the sickness is life itself. Attaining the alleged resulting will-less blissfulness in life is preferable to dying unenlightened. This is further confirmed by Schopenhauer's noteworthy *approval* of a particular kind of suicide, namely a passive "death by voluntary starvation that emerges at the highest levels of asceticism" in which "the complete negation of the will can reach the point where even the will needed to maintain the vegetative functions of the body through nutrition can fall away" (*WWR* 1, 428). The total ascetic, or life-denying "saint", may achieve such a radical lack of concern for self-preservation, and for him or her the state of will-lessness would matter far more than the choice of death. Thus the contradiction that we suspected in Schopenhauer's position does not really arise. It is coherent to say both that death is always to be welcomed as a release from life, and that releasing oneself from life while fixated on the attainment or non-attainment of individual happiness is a failure to attain the highest good.

However, this last case, that of self-induced death through ascetic abstinence, raises other alarm bells, not only about its desirability, but about its coherence within Schopenhauer's account. Jacquette pursues other contradictions here: "for a subject to have any sort of preference about living or dying contradicts what is supposed to be the saint's absolute indifference to life and death" (Jacquette 2000: 54). Death cannot be something we want or approve of, if we are truly will-less. If we assume that the enlightened ascetic or "saint" has already gained the "liberation" that consists in negation of the will, then he or she cannot *prefer* death as a release, and so cannot coherently be described as dying voluntarily. In response, however, we may point out that Schopenhauer regards the saint as having to struggle to maintain will-lessness. An absolutely stable ascetic saint (could such a being exist) might indeed attain a complete indifference between living and dying. For him or her the state of complete indifference would be the good, and death through cessation of willing would be merely a by-product that gave evidence of that good's having been attained. But Schopenhauer seems to acknowledge that this extreme state is not really possible of attainment:

> the peace and blissfulness we have described in the lives of saintly people is only a flower that emerges from the constant overcoming of the will, and we see the constant struggle with the will to life as the soil from which it arises .... Thus we also see people who have succeeded at some point in negating the will bend all their might to hold to this path by wresting renunciations of every sort from themselves, by adopting a difficult, penitent way of life and seeking out everything they find unpleasant: anything in order to subdue the will that will always strive anew. (*WWR* 1, 418–19)

While this seems to posit a distinct kind of willing that desires to be free of the will to life, (thereby shifting the lurking suspicion of contradiction now on to the notion of a "complete" will-lessness), it does at least suggest that life is always going to involve some kind of struggle, either to satisfy one's egoistic will, or to lose it. And from either kind of struggle death can still be a welcome release.

3 Epicurean arguments

So far we have assumed that for Schopenhauer, given his view that life is essentially composed of suffering and otherwise empty of positive value, non-existence would be a kind of positive good, though not the highest good. But there is a possible alternative to this position, namely that death, assuming that it is non-existence, can be neither good nor bad for the person who dies, whoever they are. Schopenhauer expounds some oft-expressed Epicurean arguments, the argument from the symmetry of non-existence post-death and pre-birth — "It is irrefutably certain that non-being after death can be no different from non-being prior to birth, and therefore no more lamentable. An entire infinity has passed when we did not yet exist: but this does not upset us in the least" (*WWR* 2, 483) — and the argument from consciousness:

> it is in and of itself absurd to consider non-being an evil since every evil, like every good, presupposes existence and indeed consciousness; and consciousness comes to an end when life ends, as it does in sleep or

> in a faint; and so the absence of consciousness is no evil .... Epicurus considered death from this perspective and thus said quite rightly ὁ θάνατος μηδὲν πρὸς ἡμᾶς (death is nothing to us); with the explanation that when we are, death is not, and when death is, we are not .... It is clearly no evil to lose what we cannot miss: and so passing into non-existence ought to trouble us no more than not having been. So from the standpoint of cognition there does not appear to be any reason at all to fear death: but consciousness consists in cognition; and so death is no evil for cognition. (*WWR* 2, 484–5)

The emphasis here is on "no evil"; but note that Schopenhauer also inserts the corollary "like every good". The upshot should be that for any time at which we do not exist, the state of affairs pertaining then can be neither good nor bad for us. And it cannot be rational to fear a state that can be neither good nor bad for us.

However, this Epicurean position does not provide a stable conclusion for Schopenhauer. There are two reasons for this. First, he thinks that the rationality or otherwise of fearing a state in which we do not exist could be decisive for us only if we were pure, rational intellects. As he explains,

> in fact it is not this cognitive part of our I that fears death — rather, the flight from death that fills all living things comes only from the blind will. But, as mentioned above, this flight is essential to the will, just because it is the will to life, and its entire essence consists in the urge to life and existence (*WWR* 2, 485)

The self is complex: we are, in one aspect or ourselves, a "pure subject of cognition" that passively witnesses the world, and can look upon the world as a "cold and disinterested spectator" (*WWR* 2, 291). If the "I" were only that conscious spectator, its own non-existence would not matter to it. But the "I" also comprises the living, willing individual, in whom the striving for life and its apparent goods is endemic, and this more fundamental part of the self will not so easily be persuaded out of its aversion to life's ending.

Secondly, Schopenhauer's position contains an implicit rejection of the Epicurean view, since he says not that non-existence is neither good nor bad, but rather every individuality is only a special error, a misstep, something that would be better off not being, and in fact the true purpose of life is to retrieve us from it' (*WWR* 2, 508). This line of thought shares with the Epicurean arguments an opposition to the attitude of the naïve, will-driven human individual, but opposes it in a different way. For if not being is an attainable purpose that is better than the conscious state in which things can be both good and bad for us, then not being is after all some kind of good. But that in turn means that something can be good for someone independently of their being conscious of it and of their existing. So, once again, the position seems to be that my death is my non-existence, and that it is a good, because existence is basically flawed and undesirable.

4 Continued existence

Despite all of the above, Schopenhauer ultimately wants to claim that my death is not my absolute non-existence. By his lights I should believe that something about me is indestructible. Not, as we said above, anything like a soul or immaterial substance, which Schopenhauer wholly rejects (see *WWR* 1, 519–21; *WWR* 2, 213) — but then what about me is supposed to be indestructible? First, Schopenhauer rather eloquently presents a kind of materialist version of continuity after death:

> 'What?' people will say, 'the permanence of mere dust, of crude matter, is to be regarded as a continuation of our being?' — But oh! Are you acquainted with this dust? Do you know what it is and what it can do? Get to know it before you despise it. This matter that lies there now as dust and ashes will, when dissolved in water, sprout into a crystal, glisten as a metal, and then electric sparks will fly from it; by means of its galvanic tension it will express a force that dissolves the most solid ties and reduces earth to metal: in fact, it will form itself into plants and animals and from its mysterious womb develop the very life that you, in your narrow-mindedness, are so worried about losing. Is it really so meaningless to continue to exist as this kind of matter? I am serious when I say that even this permanence of matter bears witness to the indestructibility of our true being, even if only in images and metaphors, or rather only in silhouette. (*WWR* 2, 489)

Yet, powerful though this picture of material continuity may be, Schopenhauer nonetheless cannot accept it as anything more than an image, metaphor, or silhouette. The reason for this is that Schopenhauer's true solution to the problem of death is metaphysical, concerning not the continuity of anything empirical in space and time, but the timelessness of the thing in itself. The materialist continuity picture appeals to him because he holds that all of the material entities that exist at any time and place are manifestations of the same essence and that this essence is "our true being". According to transcendental idealism, which Schopenhauer accepts, there is a sharp contrast between representation (*Vorstellung*) and thing in itself. Space and time, and thus matter which occupies space and time, pertain only to the world as representation, not to the thing in itself. But we can aspire to a "higher perspective" beyond empirical description that allows us to consider reality as unchanging and non-individuated. In ultimate reality there are no distinct things, and I am but a transient manifestation of the One that cannot be destroyed. Nothing that really exists can truly go out of existence, or indeed come into existence. From that perspective, Schopenhauer claims, we can assert both that the individual ceases to exist, and that the individual's true essence is not destroyed. Something timeless manifests itself for a while in the individual that I am, and it does not perish with the ending of this individual. It is in this sense that Schopenhauer can come to regard death as "unreal".

5 "I, I, I, want existence"

From the discussion so far, it looks as though Schopenhauer wants to say both

(1) At death I cease to exist

and

(2) At death I continue to exist.

Moreover, it looks as though both (1) and (2) are associated with some form of consolation. To see how Schopenhauer holds that (1) and (2) may both be true, we must recognize that he regards "I" as ambiguous: "the word 'I' contains a huge equivocation . . . Depending on how I understand this word, I can say: 'death is my total end', but also: 'my personal appearance is just as small a part of my true being as I am an infinitely small part of the world"' (*WWR* 2, 507). Someone taking the latter view could happily "leave his individuality behind, smile at the tenacity of his attachment to it and say: 'why do I care about the loss of this individuality since I carry in myself the possibility of countless individualities?"' (*WWR* 2, 507). In another passage we read "Dying is the moment of that liberation from the one-sidedness of an individuality that does not constitute the innermost kernel of our essence, but should rather be viewed as a kind of straying from our essence . . . it is this moment that can be regarded, in the sense described, as a 'restitution of wholeness'." (*WWR* 2, 524) If one can thus revise one's understanding of "I", one may arrive at an attitude to death that contrasts with that of natural, unenlightened individual.

In the second volume of his collected essays *Parerga and Paralipomena* Schopenhauer dramatizes this contrast, constructing a dialogue between Philalethes (lover of truth), and Thrasymachus, who appears on temporary loan, along with his truculent manner, from the *Republic*. Thrasymachus hears this message:

> You, as an individual, end with your death. Only the individual is not your true and ultimate essence .... Your essence in itself ... knows neither time, nor beginning, nor end, nor the bounds of a given individuality, but exists in everyone and everything. Therefore in the first sense you become nothing through your death; in the second you are and remain everything (*PP* 2, 252).

But Thrasymachus will not be consoled by the Schopenhauerian thought that, although the human individual he identifies himself as dies, the essence that he truly is (the will) exists timelessly. He replies, "Look, be that as it may, it is my individuality and that is what I am ... I, I, I, want existence. That is what I care about, and not some existence for which I first have to figure out that it is mine" (*PP* 2, 253). Philalethes then replies:

> That which cries out "I, I, I want existence" is not you alone, but everything, absolutely everything that has even a trace of consciousness. Consequently this wish in you is exactly what is not individual, but instead common to everyone, without distinction. It does not stem from individuality, but from existence as such; it is essential to everything that is, indeed it is that whereby it exists, and accordingly it is satisfied by existence as such, to which alone it refers, but not exclusively through any particular, individual existence. (*PP* 2, 253–4)

Thrasymachus represents the egoistic viewpoint that Schopenhauer considers natural to any cognitive being: everything centres on the well-being and existence of this individual that I am. His higher viewpoint involves the realization that one individual rather than another is of no great significance. Yes, the survival of the individual beyond death is an absurdity, and yes, the natural part of us that wills life recoils from annihilation. But, in Schopenhauer's view, if I can realize that my essence is identical with the timeless essence of the whole world, then I do not have to think of what I am as ceasing with the death of the individual. And since time is ideal rather than transcendentally real, the distinction between individuals across time dissolves:

> We sit together and talk and get excited, our eyes glow and our voices get louder; this is exactly how others sat a thousand years ago, it was the same and they are the same ones, and it will be just the same a thousand years from now. The contrivance that keeps us from realizing this is time. (*PP* 2, 249)

Schopenhauer's doctrine that individuality is illusory gains its immediate support from Kant's transcendental idealism. But he also places himself is a network of different traditions, starting with the Vedas and Upanishads, but including Eleatics, Neo-Platonists, Scotus Erigena, Sufism, Giordano Bruno, Spinoza, and Schelling (BM, 251–2). He finds the rejection of the individuated self echoed in other places too:

> In the *German Theology*... it says: "In true love there is neither I nor Me, Mine, to Me, You, Yours, and the like." ...Buddha says: "my students reject the idea 'I am this', or 'this is mine'." In general, disregarding the forms introduced by external circumstances and getting to the bottom of the matter, Shakyamuni [the Buddha] and Meister Eckhart teach the same thing. (*WWR* 2, 628–9)

Probably the most significant of these influences comes from the Upanishads, which Schopenhauer read all his life —though admittedly in an idiosyncratic version[7] — and from which he quotes one of the four Great Sayings (*Mahāvākyas*), namely *tat tvam asi*, or "that you are".[8] This gives expression to the doctrine that *ātman* is *Brahman*, the self is the Absolute. This is not a view that there is no self, rather that your self is not distinct from the Absolute. Perhaps even more pertinent is another of the Great Sayings, *aham Brahmāsmi*, "I am Brahman". At least in barest outline, this is the kind of metaphysical view Schopenhauer requires, if there is to be some sense of "I" that refers not to the individual but to something that endures all change in the existence of individuals.

6 Conclusion

In an attempt to sum up, the first obvious point to make is that Schopenhauer's view of death is far from simple, and comprises not so much a single verdict as an evolving sequence of meditations from shifting points of view. Schopenhauer's starting point is that, as an ordinary naïve human being who is essentially egoistic, my predominant tendency is to will the well-being of that one individual upon which my world is centred. Because any such being is driven by the will to life, death is something I will naturally fear. Schopenhauer is quite clear that personal immortality is not something to be hoped for: death is the annihilation of this individual and of the consciousness that this individual enjoys. A little Epicurean reflection suggests the thought that, precisely because it is accompanied by absence of consciousness, non-existence after death cannot be really bad for me. So it is irrational to fear being

---

[7] The *Oupnek'hat*, translated by Anquetil-Duperron. For an exhaustive account of this work and Schopenhauer's use of it, see App 2014.

[8] See *BM*, 254; *WWR* 2, 616; *PP* 2, 199, 336.

dead. However, that argument may not have much persuasive force, if the recoil from death does not have a rational basis in the first place. If I am someone like Schopenhauer's Thrasymachus, I will not be consoled by such thoughts about non-existence: it is precisely my continued existence as a conscious individual that I want more than anything. But the next step is to ask: Is there good reason to want a prolongation of this existence? A distinctively Schopenhauerian reflection now produces the idea that existing as a living being, inevitably plagued as such beings are by desire and suffering, is a state that ought to be regretted. Once we open ourselves to that Schopenhauerian outlook, the non-existence of this living being takes on the aspect of a liberation and a relief. Thus, while still viewing matters from the point of view of individual existence, but now with a revised assessment of its value, we may come to welcome the prospect of death, even despite continuing to fear it.

However, if that were the final resting-place for Schopenhauer's meditation, he would have no reason not to advocate suicide and would fall into the kind of contradiction Jacquette discussed. Instead, Schopenhauer adopts his "higher perspective", according to which any good there might be in the individual's non-existence, or indeed in the individual's existence, is outweighed by the value of attaining the blissful state of will-lessness. To reach detachment from desires, at least those that centre upon the individual's well-being, or upon the "I, Me, Mine" that constitute a fall into error and illusion for the Christian mystics and their counterparts in Indian thought, is the superior way to attain release, redemption, salvation from life. It is because this altered consciousness of self is the "highest good" for Schopenhauer that he can treat the individual's self-destruction while in the snares of egoistic delusion as an error. Finally, if the individual we so fondly identify with is ultimately to be recognized as an illusion, we may be open to the vision, albeit uncommon in modern Western philosophy, though widespread in the history of thought more generally, that we are not distinct from the "One and All", which is not destroyed when any one individual dies. Thus the consolation Schopenhauer offers at the end of his reflections is truly accessible only to someone who is open to such a broadly religious or mystical view of the relation of the self to the world.

References

[1] Anquetil-Duperron, Abraham Hyacinthe, Oupnek'hat (id est, secretum tegendum), vol. 1 (Argentorati: Levrault, 1801).
[2] App, Urs, Schopenhauer's Compass: An Introduction to Schopenhauer's Philosophy and its Origins (UniversityMedia, 2014).

[3] Cartwright, David E., Schopenhauer: A Biography (New York: Cambridge University Press, 2010).

[4] Jacquette, Dale, 'Schopenhauer's Circle and the Principle of Sufficient Reason', Metaphilosophy 23 (1992): 279–87.

[5] Jacquette, Dale, 'Schopenhauer on the Antipathy of Aesthetic Genius and the Charming', History of European Ideas 18 (1994): 373–85.

[6] Jacquette, Dale (ed.), Schopenhauer, Philosophy, and the Arts (Cambridge: Cambridge University Press, 1996).

[7] Jacquette, Dale, 'Schopenhauer's Metaphysics of Appearance and Will in the Philosophy of Art' in Jacquette (ed.), Schopenhauer, Philosophy, and the Arts (1996), 1–36.

[8] Jacquette, Dale, 'Schopenhauer on Death', in Christopher Janaway (ed.) The Cambridge Companion to Schopenhauer (Cambridge: Cambridge University Press, 1999), 293–317.

[9] Jacquette, Dale, 'Schopenhauer on the Ethics of Suicide', Continental Philosophy Review 33 (2000): 43–58.

[10] Jacquette, Dale, The Philosophy of Schopenhauer (Chesham: Acumen, 2005).

[11] Jacquette, Dale, 'Schopenhauer's Proof that the Thing in Itself is Will', Kantian Review 12 (2007): 76–108.

[12] Jacquette, Dale, 'Schopenhauer's Philosophy of Logic and Mathematics', in Bart Vandenabeele (ed.), A Companion to Schopenhauer (Chichester: Wiley-Blackwell, 2012), 43–59.

[13] Janaway, Christopher, 'What's So Good about Negation of the Will? Schopenhauer and the Problem of the Summum Bonum', Journal of the History of Philosophy 54 (2016): 649–70.

[14] Migotti, Mark, "Schopenhauer's Pessimism and the Unconditioned Good." Journal of the History of Philosophy 33 (1995): 643–60.

[15] Plümacher, Olga, Der Pessimismus in Vergangenheit und Gegenwart: Geschichtliches und Kritisches (Heidelberg: Georg Weiss, 1888).

[16] Reginster, Bernard, "Schopenhauer, Nietzsche, Wagner." In Bart Vandenabeele (ed.), A Companion to Schopenhauer (Chichester: Wiley-Blackwell, 2012), 349–66.

[17] Schopenhauer, Arthur, Parerga and Paralipomena, vol. 2, trans. and ed. Adrian del Caro and Christopher Janaway (Cambridge: Cambridge University Press, 2015).

[18] Schopenhauer, Arthur, Prize Essay on the Basis of Morals, in The Two Fundamental Problems of Ethics, trans. and ed. Christopher Janaway (Cambridge: Cambridge University Press, 2009), 113–258.

[19] Schopenhauer, Arthur, The World and Will and Representation, vol. 1, trans. and ed. Judith Norman, Alistair Welchman, and Christopher Janaway (Cambridge: Cambridge University Press, 2010).

[20] Schopenhauer, Arthur, The World and Will and Representation, vol. 2, trans. and ed. Judith Norman, Alistair Welchman, and Christopher Janaway (Cambridge: Cambridge University Press, 2017).

Taming the Existent Golden Mountain: the Nuclear Option

Frederick Kroon
University of Auckland, New Zealand.
`f.kroon@auckland.ac.nz`

Introduction

It is difficult to overstate Jacquette's contribution to the contemporary revival of Meinongianism. From the 1980s on he was a tireless proponent of the Meinongian project in general and of his own development of Meinongian ideas in particular. As regards the latter, he is often identified as a prominent advocate of the dual-property version of Meinongianism: the view that there are two kinds of properties, constitutive or nuclear, and extraconstitutive or extranuclear,[1] and that this distinction is enough to ward off familiar objections to Meinongianism. But while it is true that Jacquette was a tireless advocate of the dual-property version of Meinongianism, what many commentators miss is the continuing evolution of his Meinongian views, and the struggles he had in arriving at a version of his views he was completely happy with.

The present essay is a reflection on Jacquette's struggles with the way the dual-property view deals with Russell's complaint that on Meinong's view the existent round square exists, despite the fact that (as Meinong himself admits) no golden mountain exists. The objection was first presented by Russell in his famous review in *Mind* of Meinong's object theory (Russell 1905), a review that came shortly before his own theory of definite descriptions saw the light of day in the same journal. The plan of the paper is as follows. In the first section I present a brief overview of Jacquette's Meinongianism, including in particular his account of the nuclear/extranuclear distinction. The second section considers the way Jacquette uses this version of the distinction in his early work to solve the problem of the existent golden mountain (in brief, there is no such object as the existent golden

---

[1] Like most other commentators, I'll use the terms 'nuclear' and 'extranuclear' rather than the less familiar 'constitutive' and 'extraconstitutive'.

mountain, so the question of its existence is moot), and records his realisation that more is required to deal with the case of fictional objects. The section concludes with Jacquette's view that existence as a property of fictional objects within a story should be construed as a representational property: the object is *represented* by the author or in the work as existing. Section three is the section that is the most critical of Jacquette's work. It argues that this view suffers from two serious problems that threaten the stability of the view: these problems suggest that *all* properties that are correctly ascribed to fictional objects on the basis of the fictional works in which they feature should be seen as representational rather than categorical, a view that is reminiscent of a more recent form of Meinongianism, Priest's and Berto's *modal* Meinongiainism. In the final section I show why Jacquette was wise to resist such a move. Although I think Jacquette's view is in the grip of a tension, Meinongians should reject the view that what is actually true of fictional objects in virtue of the way they are represented is something ineluctably representational.

1   Jacquette's Meinongianism and Russell's objections

Jacquette's writings leave us in little doubt about his targets and his preferred version of Meinongianism. Jacquette, like Meinong, repudiates the assimilation of objects to existent objects, or, more generally, to objects that have being, an assimilation whose most influential proponent was to be Quine (Quine 1948). Numerous objects don't have any sort of being or Sein; but they can, for all that, still have properties. In particular, they can have the properties that make them what they are, and these constitute their Sosein. The golden mountain has *being a mountain* and *being golden* as its Sosein, the round square has *being round* and *being square* as its Sosein, Hamlet has *being a prince of Denmark* as part of his Sosein, and so on. None have existence as part of their Sosein, since none exist. More precisely, none of them have any form of being or Sein, where for Meinong the category of being includes, and is exhausted by, two forms of being: what he called existence (*existenz*), the mode of being appropriate to all objects that are genuinely located in space/time, as well as subsistence, the mode of being appropriate to all abstractions from the actual relationships that hold among the former. Following common practice, including Jacquette's practice, we will collapse these into a single category of existence-broadly-construed.

Crucially, Jacquette takes the properties of being golden, round, or a prince of Denmark to be very different sorts of properties from the properties of existence and non-existence. The former are nuclear properties, the latter extranuclear (see especially Jacquette 2015, Ch. 5).[2] The distinction is one that Meinong himself made in Meinong (1915), using a suggestion made by his student Ernst Mally, and is also defended in the work of Richard Routley and Terence Parsons (see Routley 1980, Parsons 1980). Existence and non-existence share their extranuclear status with such properties as the being possible, impossible, complete, and incomplete. The rough idea is that a nuclear property can belong to a Sosein, making a thing what it is, while an extranuclear property is a property that an object possesses in virtue of how it actually stands in relation to the world of objects (existent and non-existent). An object like the golden mountain doesn't exist, i.e. has the extranuclear property of non-existence, because no actual object is golden and a mountain. And because there are properties P such that the golden mountain's Sosein includes neither P nor non-P (e.g., it neither includes being *more* than 2000 metres in height nor being *no more* than 2000 metres in height), the golden mountain is also incomplete, i.e. has the extranuclear property of incompleteness. Jacquette thinks a good formal account of the distinction is that

(*) P is extranuclear iff, necessarily, for all x, x is not $P \Leftrightarrow x$ is non-P; P is nuclear otherwise (Jacquette 2015).

So much for the kinds of properties that objects can have or not have. What about the objects themselves? Early Meinong seems to have thought that *any* combination of properties determines an object, that the realm of objects is governed by Unrestricted Freedom of Assumption. But this threatens contradiction, as Russell saw with the case of the round square. The round square would have to be both round and square, and so would be a contradictory object. Meinong himself didn't see this as a deep problem, and responded that there is no reason to think that the law of non-contradiction extends beyond the actual and the possible. This didn't satisfy Russell, of course, because he understood the contradiction as follows: the round square is round and square, so both round and not round; and that is a straight-out *propositional* contradiction of the kind that the law of non-contradiction rules out. To this formulation of the problem Jacquette accepts a solution also proposed by Richard Routley: being square, the round square is indeed both round and non-round (being non-round is what Meinong calls a consecutive property, one that is entailed by an object's constitutive properties). But we can't then infer that the round square is not round since this would only follow if being non-round was extranuclear (by (*) above). The fact that the latter assumption entails a contradiction

---

[2]This chapter, entitled 'Constitutive (Nuclear) and Extraconstitutive (Extranuclear) Properties', first appeared as 'Nuclear and Extranuclear Properties' in Albertazzi et al (2001), 397-426. Page references are to the reprinted version in Jacquette (2015), 83-511.

shows that we should understand being non-round as nuclear rather than extranuclear. In short, there is no logical barrier to taking the combination of being round and square to determine an object whose Sosein includes both properties.

Russell also had another objection, the famous objection of the existent round square, which by Meinong's reasoning ought to be not only round and square but also existent (Russell 1907). (Russell used this objection to press home his worry about impossible objects, but the point is of course independent; it applies equally to objects whose nature is internally consistent, like the existent golden mountain, and that is the example I will work with.) For Jacquette, the lesson to be learned is that the Principle of Unrestricted Freedom of Assumption should be restricted to nuclear properties. That is, the appropriate Comprehension Principle says that for any set of nuclear properties there is an object that has those properties. Existence is extranuclear, so it can't be combined with other properties to constitute the Sosein of a new object, the *existent* golden mountain.

This is one of the few points where Jacquette admits to departing explicitly from Meinong. Meinong's own response to the problem of the existent golden mountain was to say that the golden mountain was existent but didn't exist. Commentators, Jacquette included, have recognised in this response Meinong's commitment to the idea of watered-down (*depotenzierte*) nuclear counterparts of extranuclear properties, where being existent is the watered-down nuclear version of the extranuclear property of existence. Meinong would similarly have said that we can coherently take the possible round square to be impossible and the complete golden mountain to be incomplete, so long as we here take 'possible' and 'complete' to stand for the nuclear weakenings of their full-strength extranuclear counterparts (possible– and complete– as opposed to possible+ and complete+, say). Meinong thought that the claim that some object is existent, possible– or complete– lacks what he called the *modal moment*, and that the modal moment was needed to ensure the thing's full-blooded existence, possibility+ or completeness+.

This is not the place to go into details about Meinong's contentious doctrine of the modal moment, the "factor", in Findlay's words,

> in which the difference between full-strength factuality and watered-down factuality consists. Full-strength factuality minus the modal moment yields watered-down factuality. Watered-down factuality plus the modal moment yields full-strength-down factuality. (Findlay 1995, 103-4)

What is worth noting is that Jacquette thinks that no such doctrine is needed to abolish the existent round square and that in any case the doctrine is independently flawed. He mounts a dilemma against the idea by focusing on the existent-plus-modal-moment round square. Meinong denies that an object can have attribution of the modal moment as part of an object's Sosein, on pain of the infinite regress that ensues if it is claimed that this property too can be watered down and included in a Sosein, with a second modal moment needed to lend the attribution of this new watered-down property full factuality (and so on): an infinite series of impotent modal moments, with any attempt at strengthening met with further weakening. In Jacquette's words, such a proposal "would afford no final characterization of factuality or real existence" (Jacquette 2015, 101). But as Jacquette also notes, if the regress is stopped at the first pass, which is what Meinong wants, why not say more simply that no object can have an extra-nuclear property as part of its Sosein, so that the attempt to add existence to the round square achieves nothing? That is the doctrine that Jacquette himself supports.

I think Jacquette was right to think the doctrine of the modal moment presents a hopeless way out of the problem. But as we'll soon see (and as Jacquette himself saw) this is not the end of the matter. There is a simple way to resurrect the problem.

2 Fiction and the problem of existence

Jacquette offers his brand of Meinongianism as a viable theory of the semantics of fiction, and not just the semantics of talk involving (apparently) non-referring descriptions like 'the golden mountain'. Other neo-Meinongians share this goal for their own versions of Meinongianism. But fiction presents supporters of the dual-property view with special problems. To see this, consider Conan Doyle's *Holmes* stories. These stories make it true that Holmes lived in London, at 221B Baker Street, and that he was a detective. But they also make it true that Holmes *existed*. The stories don't need to say this, because the point is utterly obvious: if Watson had jokingly told one of Holmes's clients 'Holmes doesn't exist', Holmes would have regarded this as the most elementary of falsehoods, far more basic a falsehood than the claim that he wasn't a clever detective. Note that the property of existence at play here is the extranuclear property of existence, not some weakened notion of existence. (In fact, it is hard to imagine what it would be like for Holmes to have only a weakened property of existence; would he appear a ghostly figure to Watson, perhaps?)

The evident falsity of the sentence 'Holmes doesn't exist' in the mouth of the person who knew Holmes best (Watson) might suggest that it is true that Holmes has the extranuclear property of existence. But that of course would be the wrong conclusion to draw, since we also think that it is categorically false that Holmes ever existed. What has gone wrong? Jacquette adopts the following way out. He thinks that while existence (the property at play in the scenario above) is indeed an extranuclear property, the relevant nuclear property that properly characterises Holmes is not existence at all but what Jacquette, following Chisholm, calls a *converse intentional* property, a property that an object is said to have in virtue of someone's psychological attitude to the object: in this case, the property of *being thought about by the author of the relevant story as existing*, or of *being described in the story as existing* (p. 106). Holmes has this property, so does Hamlet (relative to the story in which he appears), and so does every other fictional character, relative to the works in which they appear. (Jacquette motivates the strategy by considering the hypothetical case of a fictional work in which a character both *hallucinates* a snake and genuinely *sees* a snake. Both the imagined snake and the seen snake are snakes, but neither snake exists, so how is the Meinongian to distinguish them? According to Jacquette, by taking such converse intentional properties as *being hallucinated by X* to be nuclear properties. If we then consider a work in which the only thing that distinguishes two snakes is that one is described as existing, the other as not existing, we can capture their distinctness by taking one but not the other to have the property of being described in the work as existing. (See Jacquette 2015, 105-106.)

At first blush, the strategy can even be invoked to solve the problem of Watson's false claim that Holmes doesn't exist. Watson's speaks falsely in so far as Holmes has the nuclear property of being *described* by the author as existing. In short, on the strategy Jacquette is advocating purely fictional characters have all the standard sorts of nuclear properties that are ascribed to them in the works in which they appear: Holmes, say, has the nuclear properties of being a detective, living in London, being an avid pipe-smoker, having a friend called Watson, and so on. But they also have various converse intentional properties such as being *admired by Watson* in the case of Holmes. One such converse intentional property is the nuclear property of being *described* as existing, a property possessed by most but not all purely fictional objects (remember the imagined snake). But while purely fictional objects generally have this property, they never have the corresponding extranuclear property of actually existing.

What are we to make of this move? The first point to note is that it generalises. If a work of fiction ascribes any extra-nuclear property to an object that the object might well not have in reality, we should ascribe to the object the nuclear converse intentional property of being described in the work as having this property. Holmes is *represented* as complete for example: as having either P or its complement, for any property P. But it is also a familiar fact that fictional objects are incomplete: it is *not* true that for any P Holmes has P or non-P. So far so good. But there are problems. The first one is the obvious disanalogy between the properties of being admired by Watson and being described by the author as existing. The second but not the first makes reference to someone outside of the work of fiction, namely the author. The point is not simply that this individual exists. Some properties ascribed to a fictional character within a work of fiction equally involve real people who are also characters in the work of fiction (*The Day of the Jackal* describes the Jackal as someone who is feared by de Gaulle, for example). But the author herself is not like this: she stands outside the work. And even if an author, X for short, writes herself into a work of (meta-) fiction, any converse intentional properties she thereby generates need not reflect actuality. In the work, X might represent herself as an admirer of a certain character even though in real life X detests the character she has created.

This last point shows that Jacquette is simply wrong to assume that converse intentional properties are nuclear. One and the same converse intentional property might hold of a character within a work, yet not hold of a character in reality. That suggests a rather different division from the one Jacquette defends. We might say that where a converse intentional property holds of a character *from the perspective of a work* it is a nuclear property of the character, and where it holds of a character *from the perspective of reality* it is an extranuclear property of the character. But by that token, the property of being described by author X as existing is (almost always) an extranuclear property. Note that Jacquette would have rejected this way of drawing the distinction, since it effectively relies on a new distinction: that between properties holding within a work and properties holding absolutely, which he assimilates to Meinong's distinction between extranuclear properties and watered-down extranuclear properties (Jacquette 2015, 105-106).

The problem Jacquette faces can also be approached another way, one that should be familiar from the work of Ed Zalta, Graham Priest and other critics of the nuclear/extranuclear distinction. Most properties we think of as clearly nuclear seem to be existence-entailing. Try as we might, it is hard to think of anyone being a detective or a prince or an unhappy house-wife, or, simply, a *person*, without also thinking them as existing. To be a person requires an individual to be located in space and time, to be flesh and blood (so *concrete*, not abstract), and that requires existence. That is precisely the intuition we exploited when presenting the scenario of Watson's declaring Holmes not to exist. That claim struck as ludicrous because it is obvious that to be a person Holmes had to exist. If we now want to say, as Jacquette does, that existence is *not* one of Holmes's properties, even though Holmes's existence is implied by his having properties that Jacquette and other defenders of the dual-property approach regard as paradigm instances of nuclear properties, that surely threatens to doom the entire dual-property approach.

There might seem to be an obvious way out for Jacquette. He could have chosen to treat *every* property ascribed by an author to a fictional object in the same way as he treats existence. That is, he could have identified Holmes as the individual *described by Conan Doyle* as a detective who lives at 221 B Baker St., London, had a friend called Watson and an arch-enemy called Dr Moriarty, and so on. In fact, we have already seen that Jacquette counts all converse intentional properties as nuclear, so admitting such properties as nuclear properties of fictional objects is consistent with his dual-property view. Where it clashes with that view, of course, is in the further stipulation that these are the *only* properties that can rightly be ascribed to the nature of purely fictional objects — that the usual properties in terms of which fictional objects are characterised (paradigm nuclear properties like being a person, for example) should be excluded as they are existence-entailing. Since an essential part of the motivation for the dual-property view is that even non-existent objects have the properties in terms of which they are characterised, this further stipulation means that any reason to maintain a nuclear/extranuclear distinction among properties disappears.

Call properties denoted by predicates of the form 'being described / characterised / represented, etc., as P in work W/by individual X' *representational* properties. Note that these overlap but do not coincide with Jacquette's converse intentional properties; in particular, properties like being admired by Watson, or feared by de Gaulle, are converse intentional but not representational.

3 Meinongianism and the role of representational properties

For a reason to be discussed below, I think this criticism of Jacquette is not quite the end of the matter. But it does present us with a powerful prima facie reason to look elsewhere for a solution to the Russellian objections to Meinongianism that Jacquette was grappling with. What options does a Meinongian have at this point? I'll briefly consider just two, Zalta's dual-copula view and Berto and Priest's modal Meinongianism, before reflecting on what these accounts can tell us about the problems that Jacquette's solution faces.

Both positions deny that there is a viable distinction to be drawn among properties. But that requires them to offer a different model of the way non-existent objects, including fictional objects, have properties. Zalta famously offers a dual-copula approach (Zalta 1983, 1988), distinguishing exemplification (the familiar way in which entities like us have their properties) from encoding. The golden mountain, on this approach, is an abstract object that encodes being golden and a mountain, while the existent golden mountain encodes being golden and a mountain and existing. Neither object exemplifies existence. The case of fictional objects is similar. Holmes is the unique abstract object that encodes exactly the properties that Holmes exemplifies in the Conan Doyle stories, and that includes ordinary existence. As to the properties Holmes *exemplifies*, there matters are very different. Although Holmes doesn't exemplify being a detective or being existent, he does exemplify being an abstract object. In addition, he exemplifies such properties as being admired by many readers of the *Holmes* stories. He might even exemplify some properties that he also encodes, such as *having many admirers*.

Jacquette resolutely rejects Zalta's dual-copula approach (Jacquette 2015, Ch. 11), for reasons that needn't detain us. But he also thinks he can model that approach in terms of the nuclear/extranuclear distinction, and he thinks that shows that the data Zalta wants to capture in his theory can be equally well accounted for on his dual-property approach. He offers the following reduction:

$\forall x_1[x_1 F^1 \leftrightarrow [\neg E!x_1 \wedge A(F^1(!), x_1)]]$, where $A(F^n(!), x_1 \ldots x_n) =_{df}$ 'property F^1 (!) is attributed to the nature or Sosein of $x_1 \ldots x_n$'. (Jacquette 2015, 255)[3]

In short, given any putative non-existent object x, x *encodes* a property F, whether it be nuclear or extranuclear, just when the property F is attributed to the nature of object x. Note the following: on this reduction, encoding a property F, including the kind of property Jacquette calls nuclear, counts as functionally identical to having the property $\lambda x[F$ *is attributed to* $x]$, a representational property. So on Jacquette's own reduction of Zalta's scheme, the nuclear properties re-emerge as a class of representational properties. Given the problems faced by Jacquette's own version of the nuclear/extranuclear distinction, it is beginning to look as if our best hope for replacing Jacquette's account with something that retains something of the substance of that account is in terms of a distinction between representational properties and others.

Priest's and Berto's modal Meinongianism has no distinction between kinds of properties or modes of predication. Instead, its account of the nature of objects that are given in terms of descriptions or, in the case of fictional objects, stories, directly invokes representational properties. Objects are characterised in terms of certain properties — for example, an object may be represented as being a unique golden mountain or a unique round square or as the person who has all the properties that Holmes is credited with in the *Holmes* stories. Unless the object exists, it doesn't have these properties in the actual world but only in worlds in which the

---

[3] '$F^1(!)$' signifies that F^1 may or may not be extranuclear, where $F^1!$ is extranuclear and F^1 nuclear.

object is the way it is characterised to be. (Formally, this is secured by construing 'It is represented in work W that p' as a modal operator, subject to a version of [possible-] world semantics; Priest 2005). Where the properties are existence-entailing, the object also exists in those worlds. Holmes exists in worlds that realise the *Holmes* stories, and the golden mountain exists in worlds in which it is golden and a mountain (and so, of course, does the existent golden mountain). The round square, in turn, is round and square but only in impossible worlds; and because Priest takes existence to be physical existence, he thinks it doesn't exist in those worlds.

These objects also have properties that are non-representational. In particular, they have properties that reflect their relationship to things that actually exist. The golden mountain is an object that Meinong often thought about, for example, just as Holmes is one of the best-loved characters in fiction. As on Zalta's view, these properties are not regarded as in any way different in kind from ordinary characterising properties, except in so far as they are not existence-entailing. (They may even form part of the way an object is characterised. Take the object x such that x is both a golden mountain and my favourite object. This object may indeed be my favourite object; in that case it is my favourite object in the actual world, where it is an object that doesn't exist, as well as my favourite object in worlds in which it does exist and is a golden mountain.)

4 A problem

I have been arguing that there is a close connection between the properties Jacquette classes as nuclear properties of objects and representational properties. More precisely, I have been arguing that scepticism about Jacquette's nuclear/antinuclear distinction is naturally translated into a view that construes nuclear properties — the properties that specify an object's nature, and so allow an a priori determination of what the object is like — to be representational properties. (Another option is to take something like Zalta's dual-copula approach, but as I have noted Jacquette not only thinks he has independent reason to reject that approach, but also that this distinction can be reduced to a version of the distinction between representational properties and others.) Why not, then, understand the nature of Meinongian objects in terms of representational properties rather than nuclear properties? Although Jacquette never directly considered this option, there is little doubt that he would have rejected it. And the reason — at least, what I shall argue might well have been his reason, had he considered the option — is instructive, and says a great deal about the virtues of an approach like Jacquette's vis-à-vis an approach like modal Meinongianism.

Consider the following problem. Let ϕ be the big conjunctive predicate in terms of which Holmes (h) is characterised in the *Holmes* stories. So the Holmes stories represent Holmes in the following way: $\phi(h)$. According to modal Meinongianism, $\phi(h)$ is indeed the case in worlds that realise the *Holmes* stories, while the actual world is not like this: in the actual world Holmes is just *represented* as being like this. Now ask: what, in that case, is Holmes? How do we pick "him" out when declaring that "he" is the subject of this representation? (Because being male is existence-entailing, the neutral "it" would be more precise. But let us stay with "him".) The obvious answer is that we defer to Conan Doyle. Whoever Conan Doyle picked out, he is the one we have in mind. But this just postpones the problem. How does Conan Doyle do it? Not by stipulating that Holmes is the unique entity who bears the name 'Holmes', is extremely clever, became a detective, went to live with someone called Watson in 221B Baker St., London, and so on. That is simply how Holmes is *represented*. (Incidentally, this worry is independent of the modal element in modal Meinongianism. The worry arises as soon as you think that non-existents like Holmes or the round square only have representational properties as part of their nature, whether or not you also think that they inhabit other possible or impossible worlds. In short, it is a worry for all representationalist versions of Meinongianism.[4])

Note that we have no such problem where ordinary existents are concerned. Tolstoy knew of Napoleon, and represented him as being a certain kind of person when writing *War and Peace*. But Berto and Priest disagree that the cases are different. In response to an objection (Kroon 2012), they offer a revised account of Priest's earlier account of the so-called Characterisation Principle for modal Meinongianism:

> If Φ is a choice function on sets of objects, then the denotation of $\epsilon x A(x)$ is: $\Phi(\{x : (\alpha \text{ and } A(x)) \text{ or } (\neg \alpha, x \text{ is a non-existent object, and the envisaged world is one where } A(x)\})$ [where α abbreviates "At @, some y is such that $A(y)$"] (Berto-Priest 2014)

This suggests that the idea of selection-by-intention of an object as the denotation of a description $\epsilon x A(x)$ is indifferent to whether there is something at the actual world that satisfies the predicate A. If there is, an existent object is selected; if not, a non-existent object is selected, one that satisfies A in some other world(s). Berto and Priest realise that there will be sceptics. If nothing satisfies $A(x)$, the object selected is non-existent. But how does one intend a non-existent object?

One answer is that it is the very nature of intentionality to single out an

---

[4]Including, I think, the kind of view of non-existent objects that Tim Crane has defended in various places (see especially Crane 2013).

object of a certain kind, and given a bunch of objects, one can just point mentally to one, in the same way that, given a bunch of physical objects, one can point physically to one of them (indeed, an act of physical pointing presupposes the intentional act that goes along with it: otherwise, one could be pointing at many things). And if one imagines an object with certain properties then, ex hypothesi, what one does is imagining such a thing. (Berto-Priest 2014)

This may look a promising thought. It claims that to intend an object $\epsilon x A(x)$ is to imagine an object as satisfying the predicate $A(x)$. If nothing *actually* satisfies $A(x)$, one has thereby intended a non-existent object, one that only satisfies $(A(x)$ in worlds other than the actual world. Call this thesis *IAI* (the Intending-As-Imagining thesis).

But IAI should strike one as troubling. Remember that we are to take the predicate $A(x)$ in the description $\epsilon x A(x)$ as representational: an object is *represented* as satisfying $A(x)$. Now consider the sentence 'if one imagines an object with certain properties then, ex hypothesi, what one does is imagining such a thing'. Presumably the phrase 'such a thing' stands for the intended object, the object that in the actual world is *represented* as having certain properties. Call this the *de re* reading of the term. For this object to be the appropriate target for our imaginative exercise, it must also be the object that features in the antecedent; it must be the focus of our imagining when we "imagine an object with certain properties". But what is it to imagine an object as satisfying $A(x)$, say? Here the natural reading is *de dicto*. One simply imagines a situation in which there is an object that satisfies $A(x)$, and then one focuses on this object, all within the scope of the imagining. It is not *de re* to apprehend something which one then imagines as satisfying $A(x)$, presumably by transporting oneself imaginatively to a world where it does satisfy $A(x)$. Indeed, how would one do the *de re* apprehending of the object in the first place? [5]

The alternative way to understand the argument is as follows. Suppose one goes through the imaginative exercise under its *de dicto* construal, imagining that there

---

[5]One reason for thinking that we do have such a *de re* grasp on a non-existent object is that the invitation to imagine typically involves names or definite descriptions: we are asked to imagine *Holmes*, for example, or *the* golden mountain. But nothing can be concluded from such formulations. They only show, I think, that we are invited to imagine that there is a unique object of a certain kind, as in a game of make-believe where we invite fellow game-players to imagine that "that lion hiding behind that tree has noticed us", when there is nothing at all behind the tree. A rather different problem that affects this first formulation of the argument is that, even if it succeeds in establishing that we have a *de re* grasp on an object that we then represent in the imagination as having the desired properties, it cannot rule out the possibility that what is apprehended turns out to be an *existent* object, one which is represented as having properties that don't hold of it in the actual world (cf. Kroon 2012, Berto and Priest 2014, Priest 2016).

is an object that has certain properties. By doing this, so the present construal of the argument goes, one thereby intends an object. But this object need not have the properties in question in the actual world, since there may be nothing in the actual world that has these properties: there is no golden mountain in the actual world, for example, so in imagining a golden mountain, or the golden mountain, we are imagining something that doesn't exist. All we can say, therefore, is that the object is one that is imaginatively *represented* as having the properties.

On this account of the argument, the imagining and the intending can't really be separated. But there is an obvious problem with this way of construing the argument: it is hard to see how one gets beyond a premise about the composition of the world that is being imagined to a conclusion about the actual world. Why see this imaginative exercise — imagining a world in which there is something that is golden and a mountain, say — as testament to there being something in the actual world that is being represented in the imagination as being golden and a mountain? And why one thing rather than another? Moving in this way from the world of the imagination to the actual world seems to involve the clearest of modal fallacies.[6]

5 Jacquette redux?

To avoid committing such a fallacy, it seems that the best hope for the modal Meinongian is to show how to make sense of a *de re* grasp of an object that can then be identified, on principled grounds, as being a non-existent object. What modal Meinongianism thus needs is a solution to what we might call the Identification Problem for non-existent objects. But it fails to provide a solution. All it can say about the allegedly non-existent objects in question is that they are *represented* as having certain properties, and that is no identification at all — we need to know *what* is being represented.[7] In my view, this is one of modal Meinongianism's most signal weaknesses. Contrast this with the kind of story Jacquette has on offer. For Jacquette, a Meinongian non-existent object is constituted by its nuclear nature: its having the nuclear properties of being golden and a mountain, say. We single out the golden mountain by fixing on the unique object that has these nuclear properties, an object whose distinctness from the round square and Holmes is grounded

---

[6]Needless to say, one should be careful about attributing a modal fallacy of this kind, given the identity of the founders of modal Meinongianism! For further articulation of their argument see Berto and Priest (2014) and Priest (2016).

[7]The mistake here is akin to someone's thinking they have identified an object when they think of it as the object they are referring to with the name 'Smith', say. That is no identification at all, as Kripke reminds us when articulating his non-circularity condition on descriptivist accounts of reference (Kripke 1980, 68-70).

in the difference between their nuclear properties. These objects fail to exist precisely because there is no existent object that fits the associated clusters of nuclear properties.

This solution to the Identification Problem is not perfect, to say the least. To be convincing, it needs a viable distinction between kinds of properties, and for the reasons canvassed earlier I believe it is impossible for a Meinongian theory to deliver a viable form of such a distinction. If there is a right form of Meinongianism, it is not, I am confident, to be found in the dual-property approach. But that doesn't affect the point I am making: Jacquette's version of the dual-property approach yields an answer to a vital question that representationalist forms of Meinongianism are simply unable to answer.[8]

Earlier I suggested that Jacquette continued to think about these questions to the end. I'll conclude with a brief account of perhaps his final statement on the matter, in a paper entitled 'Domain Comprehension in Meinongian Object Theory' (Jacquette 2015, Ch 4).[9] In this paper Jacquette declares that the set of intensional identity conditions of a Meinongian object — its Sosein — can contain extranuclear properties, but that only the nuclear properties in this set can be ascribed to the object as a matter of logic. As a result, the existent round square should be construed as being distinct from the round square, even though neither is existent. Similarly, a fiction in which one villain is said to exist and another villain not to exist involves two fictional objects, one of which has existence and the other non-existence among its intensional identity conditions (156). Gone is the idea that it is only the representational property of being *described* as existing that helps to fix the identity of the first fictional object. Indeed, Jacquette now claims that converse intentional properties should be classed as extranuclear rather than nuclear (158).

This new theory is superior to the old in significant respects.[10] It allows that

---

[8] The Identification Problem is also one of the targets of Bueno and Zalta (forthcoming). In defending Zalta's abstract object theory against the criticisms of Francesco Berto, the paper stresses that one of the virtues of Zalta's theory is its ability to identify non-existent objects in a way that eludes a view like Berto's and Priest's modal Meinongianism.

[9] This chapter first appeared in Seron *et al* (2015), 101–122. Page references are to the reprinted version in Jacquette (2015). Note that Jacquette nowhere writes that the view in this chapter is his new considered view. Indeed, it is easy to read the paper as yet further speculation on problems facing his and other versions of Meinongianism. More generally, Jacquette's final book-length work on Meinong, his comprehensive collection of papers *Meinong: the Shepherd of Nonbeing* (Jacquette 2015), fails to give a clear statement of which papers represent his definitive views at the time of publication. Readers in the end have been left to make up their own minds.

[10] Jacquette even allows a sense in which the existent golden mountain is existent, although to do this he invokes the idea of the modal moment: the existent golden mountain is existent-without-the-modal-moment, but not existent-with-the-modal-moment (Jacquette 2015, 157). The return of the modal moment is, I suggest, not one the more attractive features of the new theory.

existence is part of what constitutes Holmes, just as it makes the existent golden mountain the object it is, although we still can't derive the existence of either. And the theory retains its distinctive solution to the problem of identification that in my view stymies representational forms of Meinongianism. This is surely all to the good. Of course, from the perspective of the arguments offered earlier, the theory is still fundamentally flawed, since it retains the nuclear/antinuclear distinction. What this new work does show, however, is that Jacquette was not inclined to stand still. He continued to the end to worry about the problem of the existent golden mountain, and it is gratifying to see that this final bout of speculation resulted in what is in some ways a better theory. One can't help wondering what Jacquette's final considered view would have been, had he lived.

References

[1] Albertazzi, L., D. Jacquette and R. Poli (eds.). 2001. *The School of Alexius Meinong.* Ashgate.

[2] Berto, F. 2013. *Existence as a Real Property: The Ontology of Meinongianism*, Dordrecht: Springer.

[3] Berto, F. and G. Priest. 2014. ÔModal Meinongianism and Characterization: Reply to KroonÕ, *Grazer Philosophische Studien* 90: 183Ð200.

[4] Bueno, O. and E. Zalta. 2017. ÔObject Theory and Modal MeinongianismÕ, *Australasian Journal of Philosophy* 95: 761-778.?

[5] Crane, T. 2013. *The Objects of Thought*. Oxford: Oxford University Press.

[6] Findlay, J. N. 1963. *Meinong's Theory of Objects and Values*. Oxford: Clarendon Press.

[7] Jacquette, D. 2015. *Alexius Meinong, the Shepherd of Non-Being*. Dordrecht: Springer, 2015.

[8] Kripke, S. 1980. *Naming and Necessity*, Cambridge Mass.: Harvard University Press.

[9] Kroon, F. 2012. "Characterization and Existence in Modal Meinongianism", *Grazer Philosophische Studien* 86: 23-34.

[10] Meinong, A. 1960. 'The Theory of Objects', translation of 'Uber Gegestandstheorie' by I. Levi, D.B Terrell and R.M. Chisholm. In Chisholm, ed., *Realism and the Background of Phenomenology*. Glencoe: The Free Press.

[11] Meinong, A. 1915. *Über Moglichkeit und Wahrsheinlichtkeit*. Leipzig: Verlag von Johann Abrosius Barth.

[12] Parsons, T. 1980. *Nonexistent Objects*. New Haven: Yale University Press.

[13] Priest, G. 2016. *Towards Non-Being*. Second edition (first published in 2005). Oxford: Oxford University Press.

[14] Quine, W.V.O. 1948. 'On What There is', *The Review of Metaphysics* 2(5): 21-38.

[15] Routley, R. 1980. *Exploring Meinong's Jungle and Beyond*. Canberra: Australian National University.

[16] Russell, B. 1905a. 'On Denoting', *Mind* 14: 479–493.

[17] Russell, B. 1905b. 'Review of: A. Meinong, Untersuchungen zur Gegenstandtheorie und Psychologie'. *Mind* 14: 530-538.

[18] Russell, B. (1907). 'Review of A. Meinong, Uber die Stellung der Gegenstandtheorie im System der Wissenschaften'. *Mind* 16: 436-439.

[19] Seron, D., S. Richard and B. Leclercq (eds.). 2015. *Objects and Pseudo-Objects: Ontological Deserts and Jungles from Brentano to Carnap*. De Gruyter.

[20] Zalta, E. 1983. *Abstract Objects: An Introduction to Axiomatic Metaphysics*. Dordrecht: D. Reidel.

[21] Zalta, E. 1988. *Intensional Logic and the Metaphysics of Intentionality*, Cambridge, MA: MIT Press.

Reminiscences of Alonzo Church

Nicholas Rescher
University of Pittsburgh, USA.
rescher@pitt.edu

Alonzo Church (1903–1995) was incontestably one of the principal figures in the development of modern symbolic logic — an investigator whose work contributed massively and constructively to the formation of the field. His contributions were substantial in substance and far-reaching in influence. Seeing that Dale Jacquette and I were joined in our deep appreciation of the man and his work, it is thus perhaps not amiss that on this occasion for me to give some account of my own interaction with Church.

Church was the scion of a prominent Southern family — his grandfather had been librarian of the U.S. Senate, and his great-grandfather served for thirty years as the sixth president of the University of Georgia in Athens where Church Street is named after him. Already a creative mathematician as a Princeton undergraduate, Church had a brilliant career there until his age-mandated retirement in 1973 led to the launching into a second career phase at U.C.L.A. lasting over 30 years. His contributions to mathematical undecidability, his invention of the lambda calculus, and his role in developing the Church-Turing thesis mark him as a dominant contributor to the field. His 1956 Introduction to Mathematical Logic, afforded a monumental synthesis of the field as it then stood. And his further contributions served to give the field the shape that it has through his editing of the *Journal of Symbolic Logic* (JSL), his canonical bibliography of the early field, and his magisterial editing of the Reviews section of the JSL, which gave an ongoing overview of the whole realm of relevant developments. His historical sensibilities were highly developed, and at a time when no-one then active in mathematics or philosophy paid attention to the work of Gottlob Frege, Church steadily stressed its significance until ultimately his efforts led to a surge that elevated Frege's reputation as a modern master. And beyond all this, Church's many years of teaching recruited to the field a series of brilliant investigators whose contributions served to make 20th Century mathematical logic the impressive enterprise it became.

Church's courses in Princeton's mathematics department, also drew graduate students from the department of philosophy, myself among them. My contact with Church began at the midpoint of his long and productive life (1903-1995). When

I arrived in Princeton in the autumn of 1949 I had already been in contact with Church. For late in my undergraduate years at New York's Queens College I had submitted to him, as editor of the *Journal of Symbolic Logic*, a brief note on a point of many-valued logic. In response Church replied that the matter was not of sufficient interest for an independent publication but that the point was worth making and could be made in the course of a review in the JSL. I eagerly accepted the suggestion, and so my first academic publication was mediated by Alonzo Church.

When I went to visit Church in his Fine Hall Office, he received me cordially and suggestions that I might write further reviews of the JSL. And so over the next several years I became an informal assistant to Church, writing some 40 reviews over the next decade.[1] I also took several of Church's logic courses. And while it cannot be said that I am a "Church student" — that is, someone whose doctoral dissertation was written under his supervision — it would not be inaccurate to say that I was a Church protégé.

I often visited Church's office to get further assignments or bring him material. The office was a mass of books and papers bur not a mess — everything was ordered in ways no doubt known to Church alone. His way of speaking was highly characteristic somewhat slow and precise, uttered with soft southerly intonation. His discussion was efficient and business-like, without pleasantries or small talk. I cannot recall any instance when he manifested interest in my persona and doings — or indeed in anything apart for the task at hand. But in matters relating to his field his range of information was unusually far-reaching and his statements invariably careful and accurate.

And this sort of approach characterized his teaching as well. It was dedicated to the proving of significant relationship and his presentation left nothing to the imagination. Blackboard upon blackboard was filled with his large and incredibly precise writing with each i dotted and t crossed and every step along the way set out in elaborately accurate detail.

The exposition was slow and careful (and occasionally somewhat boring) but it was totally accurate and correct, marked by a conscientious care that was admirable although occasionally and somewhat frustrating to the recipient. But what drew students to Church was not his charisma as a teacher but the inherent interest of the material he dealt with and the insightful and committed manner of his dealing with it. Church was deeply concerned always to get things just right. For years an MS of his classic Introduction was positioned in the Mathematics Library with an

---

[1] Church always vetted my reviews carefully and sometimes made editorial changes for the better. But in only one case was the change sufficiently substantive that we agreed Church would be listed as co-author. (This was a review of a paper on independence proofs by Thoralf Skolem: *Journal of Symbolic Logic*, vol. 18, 1953.).

open invitation to all concerned to identify errors of commission or omission.

I came away from my various encounters with Church not only with increased information about the technicalities of logic, but with a deeper understanding of how scholarly work should be conducted. In particular I carried away various valuable lessons. One of them relates to the virtue of detailed accuracy, of endeavoring to say exactly what one means and precisely why it is that one sees this claim as warranted. Another is the utility of contextualization, of fitting one's claims into a larger setting of issue-illuminating considerations. Yet another is the fertility of collaboration since in working together one mind stimulates the other. But above all, Church was a role model for any young scientist or scholar in making manifest the merit of a sound work ethic and the value of persistent effort. I do not know whence Dale — who also manifested this virtue in spades — got his inspiration for it, but I believe that I myself in some measure derived it from Alonzo Church.

Noneism and Allism on the Objects of Thought

Tom Schoonen and Franz Berto
Institute for Logic, Language, and Computation, University of Amsterdam, The Netherlands
T.Schoonen@uva.nl, F.Berto@uva.nl

Abstract

Noneism is a version of Meinongianism, the view that some things do not exist. Allism is the view that everything exists, including those things that the noneist takes as non-existent. Since [23], there has been a discussion on whether or not one can translate the noneist theory into the allist theory and if, in that case, the differences between the two remain substantive. In this paper we propose a notion we call *Theoretical Equivalence*: two theories are theoretically equivalent, relative to an explanandum, if the models they produce to explain it are isomorphic. We take intentional objects – which are often considered as providing a *prima facie* motivation for Meinongianism – as our explanandum. We argue that noneism and allism are theoretically equivalent with respect to the problem of intentional objects, lending some support to Woodward's [40] translation of the noneist's 'to exist' into the allist's 'to be actually concrete', in the face of recent objections by Priest [31]. We also claim, however, that while in a sense this makes the disagreement between noneism and allism insubstantial, in another sense, it doesn't.

> The domain or universe of discourse includes a wide variety of objects, existent as well as incomplete and impossible nonexistent objects.
>
> Jacquette [20, p. 101]

A version of this paper was presented at the Logic of Conceivability seminar; thanks to the LoC group for their feedback. In particular, thanks to Arianna Betti, Ilaria Canavotto, Manuel Gustavo Isaac, and especially to Richard Woodward, whose comments have been very helpful.

1 Intentionality and Intentional Objects

As you read this article, there are many things you may start thinking about. You may think or worry about the theses of this paper; you may get distracted and start thinking about the dishes that still need to be washed; you may then think about Sherlock Holmes, the main character of some Conan Doyle book on your bedside table; you might even think about a round square – a weird thing some of the protagonists of the debate we will describe below nevertheless accept in their ontologies. Similarly, we, the authors, think about Jacquette's form of Meinongianism and wonder about the metaphysics of non-existent objects. This feature of mental acts – that they are about things, the directedness of thoughts – is called *intentionality*. Gallagher and Zahavi [17] characterise intentionality as follows:

> '[I]ntentionality' is a generic term for the pointing-beyond-itself proper [...] (from the Latin *intendere*, which means to aim in a particular direction, similar to drawing and aiming a bow at a target). Intentionality has to do with the directedness or *of*-ness or aboutness of [mental acts], i.e., with the fact that when one perceives or judges or feels or thinks, one's mental state is about or of something. (p. 109, original emphases)

How to define intentionality exactly is a vexed question, one we do not hope to settle here. However, we will assume some distinctions in order to focus our discussion (here we follow current accounts of objects of thought, e.g., [11, 31]).

First of all, one might argue that there are *object-directed* mental acts as well as acts that lack this object-directedness (we will leave aside whether these can be properly called 'mental acts'). For example, feelings of nausea seem to lack the kind of object-directedness that searching for something or thinking of something do have. For the purposes of this paper we will assume that thinking of and searching for something are examples of *object-directed mental acts* and we will focus on these. Secondly, we will focus on the notion of intentionality as it is used in the tradition of Brentano, Twardowski, and Meinong. That is, we will assume the 'ordinary relation' interpretation of intentionality, i.e., that it is a dyadic relation that presupposes two relata (see [17] for another possible interpretation). Finally, one may distinguish between the content and the object of thought. So, when Jacquette thought of unicorns, the *content* of this thought was very existent, whereas the *objects* of the thought are non-existent.[1]

---

[1] The distinction between the object and content of thought was first made by Brentano's student Twardowski [37]. Betti clearly formulates the contribution of Twardowski as follows: "He [i.e.,

Our paper will focus on the debate about *the nature* of these objects of intentionality. In this sense, our discussion is located in metaphysics as opposed to the philosophy of mind (which, arguably, concerns the *content* of intentionality). Below, we will briefly discuss two contemporary accounts of the metaphysics of intentional objects and work towards the main hypothesis of our paper.

1.1 The Metaphysics of Intentional Objects

In the Meinongian tradition of intentionality, mental acts are directed towards an object, i.e., you think *about* something, e.g., when you read this article. However, it is unclear *what* these objects are, especially when our mental acts are directed towards objects that, at least *prima facie* or seemingly, do not exist. For the obvious question then arises, what is it, if anything, that our mental act is directed towards?

We will assume that we can have mental acts, like that of thinking about Sherlock Holmes, or about some dragon, that are directed towards seemingly non-existent objects. Although this is rather controversial, there are many who agree with us (amongst others [37, 26, 34, 20, 11, 31]). Moreover, we take it that there is rather intuitive evidence for it as well. Consider the following example of Priest:

(1) I thought of something I would like to give you as a Christmas present, but I couldn't get it for you because it doesn't exist. [31, p. 152, fn. 25]

This is a very mundane, everyday, thought and similar thoughts occur on a regular basis in everyday life. For example, we can think of Vulcan, Santa Claus, Sherlock Holmes, a dragon, and much more. Relatedly, consider the apparent similarity of the following truths:

(2) Some kings of England died violently and some did not.
(3) Some characters in the Bible existed and some did not. [11, p. 17]

We take (1) and (3) to be *prima facie* evidence that mental acts can be directed towards objects that, seemingly, do not exist; and that true things can be said about them. These are data a theory of intentional objects needs to account for.

So, our aim here is not to engage with those who *do not* think that we can have intentionality towards things that, seemingly, do not exist, for there are no such

---

Twardowski] drew inspiration from arguments in favor of the content-object distinction present in Bolzano, but he reinterpreted them in a Brentanian framework to sustain conclusions that were opposite to Bolzano's and that were new for the Brentanians" [6, §2]. This is the starting point of and inspiration for, Meinong's [26] famous theory of objects.

things at all.[2] Rather, we want to discuss two views that both are, in some sense, realist about said objects – they accept that such things are there, and can play the role of targets of intentional acts – but propose a, seemingly, different view of the ontological status of these objects of thought.

Below, we will consider two views that can be used to give candidate accounts of such objects of thought: *allism* and *noneism*. The comparison of these two has been a hotly debated topic since Lewis' [23] seminal paper. After an initial presentation of the accounts, we aim to contribute to this debate through a certain interpretation of the differences between the two.

Noneism

Noneism is a kind of Meinongianism – broadly, the view that some objects do not exist. Dale Jacquette has played an important role in the revival of Meinongianism (cf. [20, 21]). Typically, Meinongians are not only realists on the objects of thought – they accept such things – but also, they claim that these things really – not just seemingly, or *prima facie* – do not exist: there are things in the world, which just lack the feature of existence. Meinongians hold that one can have mental states directed to objects that do not exist, and that one can make true claims on them.

The term "noneism" was introduced by Routley [33] and it is a particular form of Meinongianism: its specificity consists in holding that there is a unique sense of being or existence (arguably, this was not the view of Meinong himself). The corresponding property or feature is a "real property" in Kant's sense: a feature that some objects have, other lack.[3]

---

[2]For example, Broad [7] held that we could not. He says of a thought of dragons that, if "true, it is certain that it cannot be about dragons for there will be no such things as dragons for it to be about" (p. 182, found in [11, p. 8]).

[3]There are many subtly different forms of neo-Meinongianism and one does well to carefully distinguish between these. For example, one that is often confused with noneism is so-called *Modal Meinongianism*, however, they concern significantly different issues. Noneism holds, as said above, that there is one sense of being or existence and that there are objects that do not exist. Modal Meinongianism, on the other hand, concerns what non-existent objects there are, how these are characterised, and what properties they can have. One can be a noneist without being a modal Meinongian and *vice versa*. Meinong, for example, was neither a noneist nor a modal Meinongian, whereas Routley was a noneist who was not a modal Meinongian – although he might have been the first to have some ideas that then became embedded in the modal Meinongian view. One of us (FB), on the other hand, is a modal Meinongian who is not a noneist, and Priest is both a noneist and modal Meinongian. See [3] for an overview of different neo-Meinongian accounts.

Allism

Allism holds that all objects exist. In this, it follows, *contra* Meinong, the view of existence held by authors like Quine, or Peter van Inwagen, according to whom the meaning of existence is, essentially, captured by the quantifier. But in particular, we take allism to be the position of those who think that all the objects of thought exist – also those that, *prima facie*, do not exist, like Holmes or a dragon. The allist also subscribes, following Quine and van Inwagen, to a unique sense of existence: that is existence-as-quantification, not existence-as-a-real-property, as the noneist has it. Also, for allists, objects may be concrete or abstract – roughly: endowed with or, respectively, devoid of, the disposition to enter into causal relations, and/or endowed with or, respectively, devoid of, spatiotemporal address.

The term 'allism' was first used, as far as we know, by Lewis [23], but it is unclear if Lewis had anyone in particular in mind who defended such a view. Within the philosophy of mathematics, both Beall [1] and Priest [28, 29] have discussed forms of allism (*Really Full-Blooded Platonism* and *Paraconsistent Plenitudinous Platonism* respectively). However, it is unclear whether there are actual proponents of allism. For example, we take it that Woodward [40], whose work we will discuss below, would not consider himself an allist as he points out that "most philosophers regard allism as being crazy" (p. 183).

Allism, just as noneism, may be put forth as providing an account of intentional objects. Or so it seems. Most Platonists hold that Platonic objects really exist, they are just abstract objects outside spacetime, that we cannot causally interact with.[4] An allist is one who claims that there really are objects of thought, but they only seemingly do not exist. In fact, they do exist, but are abstract, that is, devoid of causal features, or dispositions to causal interactions, etc. So, one might worry then that an allist, who holds that all these things really exist, cannot deal with the seeming literal truth of negative existentials ('Round squares do not exist', 'Holmes does not exist'). The allist, in turn, might claim that if one translates 'to exist' into 'to be spatio-temporally located', or 'to be endowed with causal features', in *this* sense none of the Platonic entities exist. (Below, we will discuss translation issues in much more detail.) So, the allist may hold that she can account for the data of (1), via translation.

At this point, one may wonder: is there a really significant difference between the noneist and the allist? After all, for each object accepted by one to play the role

---

[4] See, for example, debates on fictional entities where Artifactualists hold that such entities are abstract and really existent objects (cf. [38, 36]). Or Platonists about mathematical objects, who hold that numbers, though abstract, really exist (e.g., see the discussion in [32]).

of an object of thought, there is a corresponding object accepted by the other, and vice versa. One calls some of them non-existent, but the other can translate such talk in terms of abstractness, or perhaps of non-concreteness. Is the disagreement merely verbal? This tangled issue is what we shall explore.

2 Theoretical Equivalence

We believe that the answer to the above question calls for a distinction of respects: the difference between allism and noneism is significant in one respect, but not very significant in another. We will propose a notion of *Theoretical Equivalence* that holds, we will then argue, between allism and noneism, and shows in which sense the difference between the two views is *not* very significant. We will add that one should not conclude from this that the disagreement between an allist and a noneist, as a consequence of this, ends up being merely verbal: mere talking past each other due to some crucial words having different meanings in the two parties' mouths. The disagreement is still substantial, but it does not lie at the level of the *models* of the two competing views.

2.1 Modal Metaphysics and Logic: an Example

To set up our account of this metaphysical debate relative to the problem of intentional objects, let us draw an analogy with the discussion of the metaphysics of possible worlds within the scope of a theory of semantics or logic.

The metaphysical status of possible worlds has been a point of controversy ever since their formal introduction by Kripke (see [13] for an overview). However, if one turns to the discussion of this debate in logic or semantics textbooks, she always finds a very indifferent or agnostic stance towards the issue. The semanticist often stresses that she is *not* doing "heavy-duty metaphysics" when she uses the "vivid [possible worlds] way to talk about these models" [35, p. 141] (see also [9, p. 207]). The reason why semanticists and logicians often hold such an indifferent stance towards this metaphysical debate is nicely captured by Fine, who notes that "[p]hilosophers have been intrigued by the ontological status of impossible worlds. Do they exist and, if they do exist, then do they have the same status as possible worlds? To my own mind, *these questions are of peripheral interest*. The central question is whether impossible worlds or the like are of any use, especially for the purposes of semantic enquiry" [16, p. 4, emphasis added].

This reminds us of the distinction between *pure semantics* and *applied semantics* (cf. [15, 10, 30, 2]). Pure semantics is what the mathematicians and logicians do when they research the mathematical structures of modal space as 'uninterpreted', mathematical formalisms, or "pieces of mathematics" [14, p. 188]. When one wants to give an intended meaning to the logical connectives and explain the representational power of these points of evaluation, one moves to applied semantics. The move to applied semantics is characterised by the fact that "a semantics gives an account of meaning *only once* the mathematical formalism of the semantics itself has been explained *in terms of concepts relating to the actual or intended use* of the sentences of the language for which the semantics is given" [10, p. 202, emphases added].

Formulating Fine's comments through the lens of this old distinction hints towards our proposal: with respect to explaining mathematical structures, the metaphysical status of the points of evaluation (worlds) is irrelevant; however, in explaining the representational power and *how* these points capture the intuitive meaning, the metaphysics of these worlds *does* matter.[5]

2.2 Theoretical Equivalence Introduced

We aim to capture and generalise the above sentiment with the notion that we will call *Theoretical Equivalence* and then use this to evaluate the apparent similarity between allism and noneism.

DEFINITION 1. **Theoretical Equivalence**
Let there be a phenomenon that needs to be explained, i.e., an explanandum, $\mathcal{E}$, and two theories, τ_1 and τ_2, that are put forth as explanantia. Then these two theories are theoretically equivalent, with respect to $\mathcal{E}$, if the respective models purportedly doing the explaining are structurally the same, i.e., there is an isomorphism from one to the other.

Two models are isomorphic when there is a one-to-one and onto structure-preserving map between the two.[6] Notice that the notion of *Theoretical Equivalence* is relative to something: to a particular explanandum.

Our thesis is the following: allism and noneism are theoretically equivalent relative to the phenomenon of intentional objects. In order to support our thesis, we

---

[5] Arguably, there is a stronger thesis that even within applied semantics the metaphysical differences do not affect the models. Or one might disagree completely if she has a different interpretation of the pure and applied semantics distinction, that is fine. This example is just for illustrative purposes and nothing hinges on it.

[6] For formal definitions of isomorphisms, see, e.g., [19] and [18].

will explore the aforementioned issue of the possibility of a translation between the noneist and allist vocabularies.[7]

Before that, let us first briefly flag some potential confusion in order to clearly delineate how to interpret our claim that allism and noneism are theoretically equivalent, namely that our thesis is not to be confused with two stronger theses: instrumentalism and epistemic structuralism. First of all, one might think that our thesis is a form of instrumentalism in that, in the case of intentional objects, one only cares about the explanatory value of these objects in the models. However, note that instrumentalism in the philosophy of science is often described as a stronger view, namely that it sets "aside the issues of objective truth and real theory-independent existence" [12, p. 204] and holds that "unobservable things have no literal meaning at all" [8]. This is not what we advocate. Our hypothesis allows for the fact that the disagreement between allism and noneism might be very real, meaningful, and substantial, as we will see. It just does not influence the models they produce relative to the theory of intentional objects.

Secondly, one might think that for *Theoretical Equivalence* one only needs to be interested in the structure of the models and that it therefore is a version of epistemic structuralism (cf. [22]). On one characterisation of epistemic structuralism, it is a view where "we put an epistemic constraint on realism to the effect that we should only commit ourselves to believing in the structural content of a theory" [22, p. 410]. However, our notion of *Theoretical Equivalence* still allows for a view where one has more knowledge of reality than only knowledge of its structure; this is something an epistemic structuralist would deny.

2.3 Allism and Noneism: Theoretically Equivalent

Our thesis is that allism and noneism are theoretically equivalent as explanantia of the phenomenon of intentional objects. To unpack our thesis, we will use the notion of *domain of discourse*. Crane uses, equivalently, 'the universe of discourse'. As he puts it, the domain of discourse contains "all items we assume or stipulate to be relevant to our *discourse*" [11, p. 38, original emphasis]. (It is unclear if the domain of discourse is context-sensitive for Crane and whether or not it is relative to a particular subject, in a Carnapian vein. For our purposes, this subtlety does not really matter.) If (1) and (3) are evidence that we can think and make true claims about things that, seemingly, do not exist, it follows, according to Crane, that the

---

[7]Note that the use of 'paraphrasing' and 'translation' might be non-standard in the context of isomorphism, however, we use this terminology in line with the translation debate in philosophy between allism and noneism (see section 3).

mere fact that something is in the domain of discourse does not tell us much about the ontological or metaphysical status of that thing.

We take it that both the noneist and the allist agree on what intentional objects there are: all those that are in the relevant domain of discourse. This points towards the first step of a putative isomorphism f between the allist and the noneist models. Each object, o, in the domain of the noneist, is mapped via the one-to-one and onto map f to some object, $f(o)$, in the domain of the allist. Moreover, we need each property P in the noneist model to be mapped to a property $f(P)$ in the allist model which applies to the correspondingly mapped objects: o has P just in case $f(o)$ has $f(P)$. The crux of the matter is the translation of 'exists' of the noneist vocabulary into the allist vocabulary, and the mapping of the corresponding property to a suitable property in the allist model. Remember that when we introduced allism and noneism this issue was already raised and an initial stab at a translation was for the allist to interpret the noneist's 'exists' as 'is concrete'. However, as we will see below, this translation breaks down in modal contexts.

We take it that there is a translation that does the job; in particular, we think that the one suggested by Woodward [40] works. To develop our claim that allism and noneism are theoretically equivalent, we will defend Woodward's translation between the noneist and allist theories against some recent objections from Priest [31].

3 The Translation Debate

The most recent contributions in the translation debate are made by Woodward [40] (in favour of the translation) and by Priest [31] (against the translation). Our main aim in this section is to reply to Priest. However, let us first briefly go over the precursor of Woodward and Priest.

Lewis [23] was the first to point out the similarities between the noneist and the allist, when responding to the revival of Meinongianism by Routley's [33] exposition of noneism. The most important point that Lewis discusses is how the allist and the noneist can best understand each other and his discussion focuses on the issue of the quantifier. Lewis has it that we *should* interpret Routley as an allist, for even "[t]hough most philosophers regard allism as being crazy, they at least find it intelligible [...] [And] better [to be] a crazy allist than a nonsensical noneist" [40, pp. 183-186]. It is not completely clear whether for Lewis a hypothetical noneist and a hypothetical allist that seemingly disagree on the notion of existence would actually have a merely verbal disagreement – they would just be talking past each other. But it seems clear that for Lewis, who was a committed Quinean on the notion of existence, translating noneism into allism meant translating a view of which one has a hard time making sense of, into one that is understandable, though it might well be plain false.

After Lewis, the discussion has moved away from questions concerning the interpretation of the quantifier, and rightly so we believe. We agree with Woodward when he says that "[e]ven though it fits the rhetoric of Routley and Priest, there is something a little odd about the Lewisian thought that that [sic] the distinction between loaded and neutral quantification is at the heart of noneism" [40, p. 185, fn. 4], what is really at stake is that "the noneist holds that only the objects in a certain restricted domain deserve to be called 'existent' whereas the [allist] thinks otherwise" [40, p. 186].[8]

In particular, the debate focused on the translation of the noneist's 'exists' into the allist's 'is concrete'. However, as Priest, in his earlier work, has argued, this translation gets the wrong results. There are "statements whose truth-value is not preserved under [this] translation" [29, p. 155].[9] Especially, in modal contexts such as (4) and (5) the translation fails:

(4) Routley existed, but he might not have done so.
(5) Routley was concrete, but he might fail to have been so.

If we take it to be necessary whether something is concrete or abstract, as Priest does, then the translated sentence (5) is false, whereas the former seems true. So, Priest concludes, the translation fails. Note that some people accept *contingently concrete* objects (cf. [25] and [39]). However, as Woodward [40] notes, we need not accept such a controversial view. The Priestian point seems to be simply based on a very robust intuition: we may characterize concreteness and lack thereof in very different ways, but it seems very commonly accepted that everything is either concrete or not, nothing can be both, *and* concreteness is part of the essential features of whatever is concrete: if you are an uncontroversially concrete thing, like a chair or a donkey or a person, there's no way *you* could have been abstract. But, of course, you are contingent existent: your parents may never have met, or your manufacturer may never have produced you.

---

[8] See also Berto on this: "[T]heir involvement [i.e., of the quantifiers] depends on their being connected, or rather not, with the only remaining item at issue between [Quineans] and [Meinongians], that predicate, 'exists' [...]. From now on I will confront Meinongianism and Quineanism, [...], as two opposite theories of the property of existence, i.e., of what the predicate 'exists' refers to" [4, pp. 240-241].

[9] From now on, we will try to distinguish between the 'pre-Woodward' Priest by quoting the first edition of *Towards Non-Being* [29] and the 'post-Woodward' Priest by quoting the second edition of his book [31].

3.1 State of the Art: Woodward versus Priest

Woodward [40], in a recent paper, responds to Priest and aims to provide a new translation scheme on behalf of the allist that does work in the face of these challenging cases. His solution is quite simple: in order to overcome issues in modal contexts one adds an 'actuality'-condition. That is, we "interpret Priest as using 'exists' to pick out those objects that are concrete *and actual*" [40, p. 188, original emphasis]. Or, in other words, we take 'to exist' to mean 'to be actually concrete' (or 'to be actual and concrete', though we do not take there to be a difference, following Woodward).[10] This translation scheme works perfectly fine with the problematic sentence of Priest, the translation of (4) now is:

(6) Routley was actually concrete, but he might fail to have been so.

Both (4) and (6) are true, for even though Routley could not have failed to be concrete, he could certainly have failed to be actual. So, it seems that this new suggestion provides a good translation scheme between the allist and the noneist, as it "does not break down in modal contexts" [40, p. 188].[11]

So, if this paraphrase indeed completes the isomorphism then we can conclude that allism and noneism are theoretically equivalent with respect to a problem of intentional objects. (As we will see below, this translation is not without objectors.) Woodward's reformulation of his argument nicely captures this:

---

[10]Note that for the formal isomorphism, 'actually concrete' should be a non-conjunctive property, e.g., $\mathcal{C}_@$. However, it seems natural to assume this to be equivalent to the conjunctive property.

[11]It has to be noted that, in 'is actually concrete', the locution 'is actual' cannot work rigidly. So, it has to work indexically, but Priest argues that this raises some issues. "If, [...], the extension of 'is actual' varies from world to world, then for any world, w, there may be things that exist, *simpliciter*, that do not exist at w" [31, p. 202, fn. 11]. However, it is unclear what Priest means with 'exists simpliciter' and how it relates to the noneist's (his) notion of 'exists'.

First of all, note that if Priest holds that there are two ways of existing, 'to exist' and ' to exist simpliciter', then he gives up one of the main points of noneism, namely that there is only one sense of being (cf. [34, 29]). Secondly, it is unclear how this is supposed to be problematic for the proposed translation. Given that noneism, as we defined it, has a unique sense of being, Priest can either hold that this is 'exist' or 'exist simpliciter', where the former seems to be existence at a world and the latter seems to be existence at the actual world ("presumably, truth at @ coincides with truth *simpliciter*" ([31, p. 202, fn. 11], original emphasis)). The former is captured by the proposed translation of Woodward, so in that case Priest's observation does not pose a problem for the translation. Thus, only in that case that Priest's unique sense of being is existence at the actual world, there is a potential problem.

If Priest really holds that existence simpliciter is the unique sense of being that noneists have, then it is no longer clear why (4) would be true for the noneist and it might be that the allist can just translate 'exists' of the noneist to 'is concrete'. We will leave this for what it is, as we do not see how Priest could have a genuine issue with Woodward's usage of 'actually' here.

> Let's spot for the moment that noneism is true. Now imagine that we rewrite our noneism theory: whereas previously we said that an object exists, we now say that an object is actually concrete, and where we previously said that an object is self-identical, we now say that an object exists. *No one seriously thinks that this relabelling exercise has changed anything*: all we've done is rewritten the theory in a different way. But our rewritten noneist theory just is allism.
>
> [40, p. 191, emphasis added]

What is important here, is the claim that the relabelling exercise does not change anything. Let us phrase Woodward's translation in terms of our *Theoretical Equivalence*: the feature of existing from the noneist's theory is mapped, while preserving the structure, to the feature of being actually concrete from the allist's theory.

In the second edition of *Towards Non-Being*, Priest [31] raises three main objections against Woodward [40], which we will dub: (i) symmetric translation; (ii) failure of translation; and (iii) unwarranted conclusion. Let us first briefly address the worry of the symmetric translation. Priest argues that a good translation is symmetric and that, hence, there being a translation *by itself* does not give any advantage to either view. We agree, but we note that this is not how Woodward intended his argument to be taken; his argument is that *if* one accepts the translation, then she has a problem maintaining that there is a substantive disagreement.[12] Woodward is *not* suggesting that the translation itself is problematic for the noneist. In the remainder of the paper, we will address the other two objections by Priest.

3.2 What it Means to Exist

Addressing Priest's 'failure of translation'-objection is a subtle issue and hinges on a very clear understanding of the role of the translation in the claim that allism and noneism are theoretically equivalent. The reason why it is important to get this clear is that it is here we diverge from Woodward [40].

The role the translation plays for us is a technical one; it is involved in the isomorphism between the *models* that are proposed as explanantia. We do not aim to say something on the folk meaning of existence, nor on the meaning of existence as used by other theorists.[13] Importantly, this also means that if the translation works in this isomorphism-sense, nothing follows on what the intuitively correct meaning

---

[12]Thanks to Richard Woodward for helping us get clear on this. Moreover, he agrees that if the allist accepts the translation *and* holds that the disagreement is substantive, she also has to address this worry.

[13]Another reason to ignore the folk meaning of such technical terms is if one believes these to

or interpretation of 'to exist' is and whether any of the two accounts has the better account of this intuitive meaning.[14] Priest, in a sense, seems to concede to this type of translation when he notes that one might always "coin a neologism, [...], to mean whatever relation it is to the concrete that is needed to make the translation manual work. [...] The word then just becomes a term of art. Moreover, it divorces the word from whatever content it would normally seem to have" [31, p. 202]. But the fact that the translation becomes divorced from the 'normal content' is not a problem for us, because as we just pointed out, with respect to *Theoretical Equivalence*, we are not concerned with capturing the intuitive meaning of 'exists'.

The simplicity of noneism, insofar as it takes 'exists' to be univocal, is one of its main selling points over, e.g., Platonism. For example, Priest notes that "the picture of reality whereby it comprises the existent, *which are concrete objects in space and time*, and, for the rest, the non-existent, has an appealing cleanness about it" [29, p. 134, emphasis added]. So, everything, from mathematical objects [29, p. 135] to fictional objects [31, p. 317], are all non-existent objects for the noeist. Similarly, Routley points out that noneism "enables [...] that what is said to exist can coincide with what really does exist, namely only certain individual objects now *located in space*" [34, p. 152, emphasis added].[15] So, it seems that Woodward's translation does capture the noneist's use of 'exists'.

Yet, despite all this, Priest disagrees with the proposed translation. He does so based on another sentence pair that he takes to be problematic for Woodward's translation:

(7)   If 3 were an actually concrete object, then it would be in space-time.

(8)   If 3 were existent, then it would be in space-time. [31, pp. 201-203]

The first sentence is, trivially, true for both the noneist and the allist; to be actual and concrete just is to be in space-time (of this world). The second, according to Priest, "is false for a noneist—and certainly not trivially true. Plato, after all, could have been right. 'Exists' does not *mean* to be in space/time. So there can be possible worlds where things exist which are not in space/time" [31, pp. 201-202, original emphasis]. Our response to this argument of Priest is two-fold: (i) first we will argue that this response is not compatible with the type of noneism we have

---

bare no relation to the 'philosopher's use' of the technical terms (cf. [5, Sec. 3]). We will not engage with this debate here.

[14]For accounts of the natural language usage of 'exists' and 'there is', see [27], [11, Ch. 2], and [31, Ch. 17].

[15]But see also Meinong [31, p. 311] and others, [24, p. 440], [20, p. 116, fn. 2], and [3]. See also the noneist's semantics for the existence predicate, e.g., [29, p. 13].

described above and (ii) then we will respond to a possible reply by Priest, namely, that we have the wrong of type of noneism in mind.

(i) When Priest claims that there may be things that exist but are not in space-time (and, we take it, have no causal powers), and supports this by referring to Plato, he seems to give up the intuitive appeal of simple noneism. It is true that Priest seems to consider a form of Platonic noneism (he says that "a noneist can certainly endorse a platonist account" of mathematical objects [29, p. 135]), but he does not seem to support this himself ("[b]ut for the noneist, a simpler view beckons" (ibid.)). And, more importantly, this is *not* the type of noneism that we referred to since the start of the paper: the type of noneism we argue to be theoretically equivalent with allism. According to the noneist, 'exists' is univocal, and what 'exists' means can be glossed metaphysically (if not defined) by saying that what exists is in space-time, and/or has causal powers. Now, it does not seem to be compatible with this noneist view to claim that it is merely contingent that existents have causal powers or space-time location. The noneist seems to be giving a general, metaphysical characterization (if not a definition) of existence in its unique sense. Then even if 'to exist' does not *mean* to be in space-time, or to have causal powers, for the genuine noneist there are no possible worlds where something exists, but lacks space-time location or causal powers. Plato could not have been right, that is: there is no possible world where he is. There might well be worlds where things exist but lack spatiotemporal location, or causal features, but these will be impossible worlds: ways things could not be. It seems, thus, that one cannot hold that there could be things which exist but are not in space-time and/or have no causal powers, and remain a genuine noneist.

(ii) Maybe then, we have described a form of noneism that Priest does not support, as opposed to a form of noneism that would make Priest's analysis of (7) and (8) correct. If that is the case, this does not show that the translation scheme is flawed, it just shows that the noneist was not clear on what meaning of 'to exist' needs to be translated. Consider the following analogy: Franz cannot provide Tom with a proper English translation of the Dutch 'gezellig', if Tom does not first tell Franz what 'gezellig' means.[16] Again, with respect to our notion of *Theoretical Equivalence*, one is not trying to come up with a translation for 'exists' that matches the folk meaning of the word. One is trying to come up with a translation that matches the property picked out by 'exists' in the noneist's model with a corresponding property in the allist's model. What is at stake here is the theoretical equivalence between allism and noneism; not who does better with respect to the

---

[16]'Gezellig' is a notoriously difficult word to translate from Dutch to English, some say it cannot be (properly) translated at all.

folk meaning of 'exists'.

3.3 Not Talking Past Each Other

The above all dealt with Priest's objection that the translation proposed by Woodward does not work. We believe that we have sufficiently weakened the objection with respect to the work translation does for the theoretical equivalence. Let us now turn to Priest's final objection against Woodward, which is aimed at the conclusion Woodward draws from his argument. That is, Priest argues that *even if* we grant that the translation scheme works, then it is still not the case that the conclusion Woodward draws is warranted. Priest uses the following example:

> Let us suppose that I believe that Nicaragua is a country in Central America, that Spanish is spoken there, and (correctly) that its capital is Managua. You believe that Honduras is a country in Central America, that Spanish is spoken there, and (incorrectly) that its capital is Managua. Neither of us has any other beliefs about Central American countries, and in all other respects our beliefs are identical. The translation from my vocabulary to yours which maps 'Nicaragua' to 'Honduras', and otherwise leaves everything unchanged, preserves things held to be true, in both directions. Must it then be the case that 'Nicaragua' in my mouth means the same as 'Honduras' in yours? Clearly not.
>
> [31, pp. 204-205]

Now, we agree with Priest on this point: it is not the case that, when there is a translation, the disagreement is not substantive. However, we do not suggest that the translation, by itself, tells us anything; it is only with respect to a particular explanandum – to which the models function as explanantia – that we may draw certain conclusions. Importantly, this allows us to agree with Priest that there *is* a distinct fact of the matter (in this case, a public meaning or a concept) the noneist and allist are disagreeing about, without weakening the force of the translation with respect to our notion of *Theoretical Equivalence*.

To see this, consider two models of Central America, both exactly matching in the cities they represent, the lakes, and the relative distance of everything to each other. That is, for all points in the one model and all relations between the points, there correspond points and relations between the same points in the other model. One model is such that the demarcated area where the point labelled 'Managua' is located, is labelled 'Honduras', whereas in the other model, this area is labelled 'Nicaragua'. Now, if the explanandum is to help navigate the agent from point (city) to point in the most efficient way, the difference between the two models is irrelevant. That is, relative to the problem of navigation the two models are theoretically equivalent. Yet, there is a distinct fact of the matter as to what the public meaning of 'Honduras' and 'Nicaragua' is.

Similarly, even though allism and noneism are theoretically equivalent with respect to the problem of intentional objects, we still agree with Priest that there may be a fact of the matter on which the noneist and the allist, meaningfully, disagree. That is, the disagreement between the allist and the noneist might still be very substantial: "[o]ntological questions are *not shallow*, insofar as they are substantive, structural questions about the nature of such [i.e., existence] property" [4, p. 242, emphasis added]. It is just that the disagreement does not prevent the models of intentional objects of the two theories from being isomorphic – i.e., allism and noneism are theoretically equivalent with respect to the explanandum of intentional objects.

Just as we agree with Priest that it is not the case that 'Nicaragua' in one's mouth means what 'Honduras' means in the other's in the example above, we do *not* think that, in the noneist-allist debate, 'exists' in one's mouth means what 'is (actually) concrete' means in the other's – so that the two parties are *merely* talking past each other. 'Exists', in particular, both in the mouth of the allist and in that of the noneist, means whatever the word means in English: meanings are public and shareable. *What* exactly the word means, is controversial. And noneism and allism come with two very different accounts of the notion of existence: the latter asserts, while the former denies, that the notion of existence is essentially captured by the quantifier. The disagreement is very real. The theoretical equivalence view is just the claim that the two theories are posited as explaining a phenomenon via models that are isomorphic to each other. If one theory has an advantage over the other, thus, this cannot be spotted by just looking at the respective models.

4 Conclusion: Theoretical Equivalence and Disagreement

Let us take stock of where we are at this point. We started out discussing the phenomenon of intentionality and intentional objects, i.e., the objects our thoughts are 'directed towards'. We assumed that there are indeed intentional objects, some of which, seemingly, do not exist. We took the main data point to be explained by a theory of intentional objects to be provided by claims like (1):

(1) I thought of something I would like to give you as a Christmas present, but I couldn't get it for you because it doesn't exist.

We considered two theories that are put forth as accounting for the phenomenon of intentional objects: noneism and allism. The former claims that some objects of thought are what they appear to be: non-existent; whereas the latter claims of all objects of thought that they exist. We looked at the thesis that allism and noneism are theoretically equivalent, relative to intentional objects as the explanandum. The most important part of the isomorphism view is connected to how the allist should interpret the noneist's 'exists' predicate. We agree with Woodward [40] that 'being actually concrete' seems to get the translation right. In order to strengthen this point, we evaluated recent arguments by Priest [31] against Woodward's translation and argued that, with respect to *Theoretical Equivalence*, these objections lose their force. Hence, we conclude that the differences between allism and noneism have little to do with the models they produce as explanantia for (1).

However, allism and noneism's theoretical equivalence with regards to intentional objects does *not* entail that their disagreement is shallow, in a different sense: the two parties are not just talking past each other. As we pointed out above, our view still agrees with Priest that there is a fact of the matter: what the *authentic* content of the notion of existence is – as expressed, typically, by the folk or common sense of verb 'to exist' in its so-called absolute uses.

References

[1] J.C. Beall. From Full Blooded Platonism to Really Full Blooded Platonism. *Philosophia Mathematica*, 7(3):322–325, 1999.

[2] F. Berto. Is Dialetheism an Idealism? The Russellian Fallacy and the Dialethist's Dilemma. *Dialectica*, 61(2):235–263, 2007.

[3] F. Berto. *Existence as a Real Property. The Ontology of Meinongianism*, volume 356 of *Synthese Library*. Springer, Dordrecht, The Netherlands, 2013.

[4] F. Berto. There is an "is" in "There is": Meinongian quantification and existence. In Alessandro Torza, editor, *Quantifiers, Quantifiers, and Quantifiers: Themes in Logic, Metaphysics, and Language*, volume 373 of *Synthese Library; Studies in Epistemology, Logic, Methodology, and Philosophy of Science*, pages 221–240. Springer International Publishing, Switzerland, 2015.

[5] A. Betti. The Naming of Facts and the Methodology of Language-Based Metaphysics. In Anne Reboul, editor, *Mind, Values, and Metaphysics. Philosophical Essays in Honor of Kevin Mulligan—Volume 1*, chapter 3, pages 35–62. Springer, 2014.

[6] A. Betti. Kazimierz Twardwoski. In E. N. Zalta, editor, *The Stanford Encyclopedia of Philosophy*. CSLI Publications, summer 2016 edition, 2016.

[7] C. D. Broad. *Religion, Philosophy and Psychical Research*. Routledge and Kega Paul, London, 1939.

[8] A. Chakravartty. Scientific realism. In E. N. Zalta, editor, *The Stanford Encyclopedia of Philosophy*. CSLI Publications, Stanford, CA., fall edition, 2015.

[9] G. Chierchia and S. McConnell-Ginet. *Meaning and Grammar: An Introduction to Semantics*. MIT Press, Cambridge, MA., 1990.

[10] B. J. Copeland. Pure Semantics and Applied Semantics. *Topoi*, 2:197–204, 1983.

[11] T. Crane. *The Objects of Thought*. Oxford University Press, Oxford, paperback edition, 2015.

[12] M. Curd, J. A. Cover, and C. Pincock, editors. *Philosophy of Science*. W. W. Norton & Company, New York, NY., second edition, 2013.

[13] J. Divers. *Possible Worlds*. Routledge, London, 2002.

[14] J. Divers. Possible-Worlds Semantics Without Possible Worlds: The Agnostic Approach. *Mind*, 115(458):187–225, 2006.

[15] M. Dummett. *Truth and Other Enigmas*. Harvard University Press, 1974.

[16] K. Fine. Constructing the Impossible. Retrieved on 19th of April, 2016 from https://www.academia.edu/11339241/Constructing_the_Impossible, 2013.

[17] S. Gallagher and D. Zahavi. *The Phenomenological Mind. An Introduction to Philosophy of Mind and Cognitivie Science*. Routledge, New York, NY., 2008.

[18] C. Glymour. Theoretical Equivalence and the Semantic View of Theories. *Philosophy of Science*, 80(2):286–297, 2013.

[19] H. Halvorson. What Scientific Theories Could Not Be. *Philosophy of Science*, 79(2):183–206, 2012.

[20] D. Jacquette. *Meinongian logic: the semantics of existence and nonexistence*. Perspektiven der analytischen Philosophie. de Gruyter, Berlin; New York, 1996.

[21] D. Jacquette. *Alexius Meinong, The Shepherd of Non-Being*, volume 360 of *Synthese Library. Studies in Epistemology, Logic, Methodology, and Philosophy of Science*. Springer, 2015.

[22] J. Ladyman. What is Structural Realism? *Studies in History and Philosophy of Science*, 29(3):409–424, 1998.

[23] D. K. Lewis. Noneism or Allism? *Mind*, 99(393):23–31, 1990.

[24] B. Linksy and E. N. Zalta. Is Lewis a Meinongian? *Australasian Journal of Philosophy*, 69(4):438–453, 1991.

[25] B. Linksy and E. N. Zalta. In Defense of the Contingently Nonconcrete. *Philosophical Studies*, 84:283–294, 1996.

[26] A. Meinong. The Theory of Objects. In Roderick M. Chisholm, editor, *Realism and the Background of Phenomenology*, pages 76–117. The Free Press, 1904. Translated by I. Levi, D. B. Terrell, and R. M. Chisholm. 1960.

[27] F. Moltmann. The semantics of existence. *Linguistics and Philosophy*, 36(1):31–63, 2013.

[28] G. Priest. Meinongianism and the philosophy of mathematics. *Philosophia Mathematica*, 11(3):3–15, 2003.

[29] G. Priest. *Towards Non-Being; the logic and metaphysics of intentionality*. Oxford University Press, Oxford, 2005.

[30] G. Priest. *Doubt Truth to be a Liar*. Oxford University Press, Oxford, 2006.

[31] G. Priest. *Towards Non-Being*. Oxford University Press, Oxford, 2nd edition, 2016.

[32] G. Restall. Just what is full-blooded Platonism? *Philosophia Mathematica*, 11(1):82–91, 2001.

[33] R. Routley. *Exploring Meinong's Jungle and Beyond*. Australian National University RSSS, Canberra, 1980.

[34] R. Routley. On What There is Not. *Philosophy and Phenomenological Research*, 43(2):151–177, 1982.

[35] T. Sider. *Logic for philosophy*. Oxford University Press, Oxford, 2010.

[36] A. L. Thomasson. *Fiction and Metaphysics*. Cambridge University Press, 1999.

[37] K. Twardowski. *On the Content and Object of Presentations*. Martinus Nijhoff, The Hague, The Netherlands, 1894. Translated and with an Introduction by R. Grossman. 1977.

[38] P. Van Inwagen. Creatures of Fiction. *American Philosophical Quarterly*, 14(4):299–308, 1978.

[39] T. Williamson. Bare Possibilia. *Erkenntnis*, 48:257–273, 1998.

[40] R. Woodward. Towards Being. *Philosophy and Phenomenological Research*, 86(1):183–193, 2013.

Why Nothing Fails to Exist

Peter Simons
Trinity College, Dublin.
psimons@tcd.ie

> a curious problem is raised by the denial
> of a singular existential statement; e.g.,
> "There is no such thing as Pegasus."
> W. V. Quine, Designation and Existence, 1939.

1 Intentionality and the non-existent

One of the many areas of Dale Jacquette's knowledge and expertise was the history of the intentionality theory that emanated from the work of Franz Brentano. His essay "Brentano's concept of intentionality", in the volume he edited, *The Cambridge Companion to Brentano*, traces a wonderfully clear path through the sometimes tortuous developments from Brentano's 1874 *Psychologie*, through the criticisms and alternatives of his students, to his final reistic theory. Sharing, like Dale, a fascination with Brentano's theory and its successors, including a more than sneaking admiration for Brentano's *bête noire* Meinong, I am pleased to return to the old conundrum of thought about the non-existent. The purpose however is not to add yet more curlicues to the historical narrative of that period, which has been intensely worked over, but to think squarely about the problem itself, and propose a view that is different from but as un-Meinongian as that of the late Brentano.

2 Thinking about what does not exist

I shall use 'exist' in the maximally neutral, non-tensed sense in which past and future objects may be said to exist just as much as present ones, in which events and processes exist just as much as substances, and in which abstract entities like universal properties, states of affairs, platonic forms and mathematical objects may be said to exist, if indeed there are such things. It is a datum of this investigation that, in some relatively untechnical sense, we may rightly be said to think (speak, dream, wonder, speculate..., but I shall use 'think' to cover all such types of act) about what does not exist. This can occur in one of two linked ways. The first is general. A scientist may postulate a kind of entity such as a kind of matter or particle, a kind of force, a species ancestral to humans, a deadly pathogen, that later research shows not to have existed. We may think of phlogiston in physics, or the four humours in medicine, Freud's ids, or ghosts and fairies. The second kind of thought about the non-existent is non-general: an individual or definite group of individuals is claimed and may indeed be widely believed to have existed, but in fact does or did not, as subsequent investigation may show. The bible's Adam and Eve, Plato's Atlantis, Geoffrey of Monmouth's King Arthur, Le Verrier's planet Vulcan, and the supposed fourth bullet of Kennedy's assassination would all be candidate examples. Clearly fictional and mythical examples would add Bellerophon and Pegasus, Holmes and Watson, Frodo and Gandalf, and many more. A clear connection between the two sorts of thought is that many of the non-existents are specified by a definite description mentioning the object's kind: person, planet, wizard etc.

3 True non-general negative existentials

The truth of *general* negative existentials need not detain us. There are no unicorns, hobbits, talking donkeys or bodies that travel faster than light. There is no reason to suppose that in correctly affirming such true negatives we are in some underhand way committed to non-existent unicorns etc. It is with the true denial of the *particular* (whether singular, plural or mass) that we need to be concerned, as the motto from Quine emphasises. The datum to be accepted and explained is that there are true non-general negative existentials, such as

>Vulcan does not exist
>King Arthur does not exist
>Frodo Baggins does not exist

For a nominalist such as myself, I would add such anti-platonist denials as

>The number 2 does not exist

as further truths, but that is a controversial area so I will confine myself to uncontroversial examples. The argument of this paper is that we can accept the datum, accept these statements and similar as true, while denying that there are any non-existent objects, and this without eliminating or paraphrasing away the singular terms 'Vulcan' etc.

4 Candidate non-existent objects

Not all putative non-existent objects have the same kind of credentials, and as a result the plausibility with which it may be held that there are such objects varies. Here 'there are' must clearly be broader than 'there exist' if this view is not to be trivially self-contradictory. Let us consider the main candidates.

4.1 Unactualized possibles

Here the datum is true negative existentials of the form 'A does not exist' where 'A' purportedly refers to a non-existing but possible object. Le Verrier might been right about Vulcan, but was not:

[1] There is no intramercurial planet

[2] Vulcan does not exist

are both true. But they might have been false. The associated true existential counterfactuals are

[3] There could have been an intramercurial planet

[4] Vulcan could have existed

and *their* false opposites are

[5] There could not have been an intramercurial planet

[6] Vulcan could not have existed

A typical and sensible reaction to the latter two statements would be: *Why not? What prevented it?* A plausible answer is: *Nothing*, it just never came about. So the truths [3] and [4] are true not for the existence of something making them true but for lack of truth-makers for their opposites [5] and [6]. We come back to missing truth-makers in Section 5.

In other cases, such as the talking donkeys of David Lewis's modal realism, the impetus comes not from science but from a semantic theory of modality. Here is not the place to enter into discussion about such theories, about which volumes have been written. The ubiquity and utility of modal talk does indeed call for an account of how it works, and there is no informed consensus. The majority of theories however do not endorse modal realism: for them, the inferences from [1] and [3] to

[7] There is an intramercurial planet that exists but is not actual

and from [2] and [4] to

> [8] Vulcan exists but is not actual

are not valid. And even if they were, [7] and [8] expressly *affirm* the existence of possible but non-actual objects. Only those Meinongian versions of modal realism which take [1]–[4] at face value would be committed to non-existent objects, as

> [9] There is a possible intramercurial planet that does not exist
>
> [10] Vulcan is a possible object that does not exist

While we have not here even attempted to refute such theories, it is plain there are many alternative ways to account for so-called unactualized possibles, so it is not unreasonable to suppose that such candidates, while stronger than some others, will not force us to accept non-existents.

4.2 Meinongian Objects

In Meinong and Twardowski, an object is anything that could in principle be thought about, whether it exists or not. In this sense all objects are potential intentional objects, and any particular object of which we truly state that it does not exist, for example Vulcan, is in fact an intentional object of some thought. Since thought can be inconsistent and incomplete and yet still be about something, Meinong accepts inconsistent and incomplete objects, which are impossible, i.e. could not possibly exist, such as the round square or the greatest prime number. These are in addition to the possible objects of 4.1. The theory of non-existent objects in Twardowski and Meinong is a product of their account of intentionality, according to which any presentational mental act has both a content and an object. In Meinong, who throroughly de-psychologised his theory and became a realist about non-existent objects, this leads, along with his rich psychology of intentional acts, to a rich ontology of objects, including states of affairs, values and desideratives. Indeed Meinong's is arguably the richest possible ontology. Nevertheless, while Russell attempted to show it is inconsistent, it is not obviously so. While paradoxes like the Liar and the Russell set pose threats to Meinong,[1] his system has not been shown to be worse off than others which attempt to tackle such paradoxes. However, let us keep our feet on the ground. Meinong's impossible objects are to be avoided if we can avoid them, and since the argument for them *in Meinong* is the same as that for unactualized possibles, namely the analysis of intentionality, if the impossibles can be avoided on that account, so can the non-existent possibles.

---

[1] Jacquette, 1982.

4.3 Purely Intentional Objects

In the wake of the rejection of Brentano's in-existence or immanence theory of intentionality, subsequent theories took different paths.[2] The Twardowski‑Meinong theory accepted that all intentional acts have objects and therefore some these do not exist. A more moderate reaction was that of the early Husserl, who accepted that all intentional acts have content, but not all need have an (external, transcendent) object or referent. When I think about Zeus, there is no Zeus in my experience, "it is in truth not really immanent or mental. But it also does not exist extramentally, it does not exist at all."[3] In this Husserl was recapitulating the view of his hero Bolzano, who states in his *Theory of Science* that there are objectless ideas, whether because the ideas are inconsistent, like *round square*, or for some other reason, like *golden mountain*.[4] Bolzano's ideas (he calls them ideas-in-themselves) are abstract, platonic entities, but someone who thinks of a round square or a golden mountain has such ideas as the content of their thought, and these ideas lack objects. By 1913, when he revised his theory of intentionality, Husserl had installed noemata and purely intentional objects as non-mental correlates of acts, but these are still distinguished from the acts' external objects. When thinking of Vulcan, my thoughts have a mental content, as well as a purely intentional object, which exists and is abstract, and yet my thoughts entirely lack an external object or referent. So while Husserl, like Bolzano, is happy to accept abstract entities, at no point is he prepared to accept non-existent ones. While I disagree with Husserl and Bolzano that there are such abstract contents — which is matter for another discussion — their commonsensical rejection of non-existent objects of thought in their analysis of intentionality is the one I endorse. Thoughts with definite content may lack referents.

4.4 Fancies

These are objects of non-veridical mental phenomena such as dream, imagination, hallucination, delusion, false memory, unfulfilled anticipation etc. I call them 'fancies' because they are not strongly tied into a sense of reality and are, because of their relative lack of intersubjectivity, not very compelling candidates for non-existent objects, though for a Meinongian they are as good as any others. They are fairly easily accommodated in the Bolzano–Husserl theory of intentionality outlined above (or a modified nominalist version as I would prefer) and will not be further

---

[2] Jacquette, 2004.
[3] Husserl, 1970, Investigation V, §11, 559.
[4] Bolzano 2014, §67, 220f.

stressed. They are here mainly for completeness and to contrast them with the other cases.

4.5 Fictions

These are objects with an anchorage in texts, films, and other intersubjectively accessible media of representation. They include creatures of (now) exposed myth such as Zeus and Pegasus, of legend such Arthur and Lohengrin, patent creations like Elizabeth Bennet, Frodo Baggins, Batman and the Dark Star, and those gods and creatures of sincere but mistaken belief systems such as folk superstitions and religions (examples deliberately not given). There is both generally and in the philosophical community a much stronger inclination to hold that there are such objects, mainly because of the stability and intersubjectivity their physical anchorage affords. Literary critics can discuss and argue about the character of King Lear just as historians do about the character of Joseph Stalin, Freud could psychoanalyse Hamlet just as he did real patients. They seem somehow more "real" than fancies, and their characters, exploits and fates can move us as much as, or more than, those of real people.

There are many theories of fiction and fictions, and not all of them require there to be non-existent objects. If a fiction is considered something mental, or something abstract, then it exists in the manner that mental or abstract things do. The sophisticated theory of Roman Ingarden regards fictions as existing objects created in time by individuals or groups, but as ontologically radically dependent or heteronomous, requiring both minds and anchoring media for their continued and derivative existence.[5] Such fictions have a double structure of properties: in itself, the fiction Frodo Baggins is a character created by J. R. R. Tolkien in the 20th century, enjoys popularity, features in films etc. But it is ascribed or imputed such properties as being a hobbit, a ringbearer and so on. So for Ingarden, Frodo exists, but he's a fictional character rather than a hobbit: there are no hobbits. Meinong's account of fictions on the other hand does require them not to exist. According to Meinong, an author does not create an object when he or she writes a story, but picks out or selects one of an infinite number of non-existing objects Ð non-existing because any fictional narrative is necessarily incomplete in its description of an object's properties.

My own nominalist account of fictions does not rely on abstract or heteronomous objects as in Husserl or Ingarden but is a social conspiracy theory. There are generally willing (sometimes erroneously believed) conspiracies of coordination of imagination, suspension of disbelief etc. among consumers of fiction, whether literary or

---

[5]Ingarden, 1974.

other. While this conspiracy can break down, as when people misinterpret or wilfully bend or augment an author's work, in those cases where it is relatively stable and harmonious, there are grounds in the author's descriptions, depictions or other renderings for connoisseurs to agree and disagree about what the fictions in question would be like if they were real. The grounds for this stability and agreement are, as Ingarden said, both the intersubjectively accessible anchors of texts, pictures etc., and the plethora of mental acts experienced by the producers and consumers of the fiction, the detailed account of which can be delegated to phenomenology. It will be complicated in detail, but in no case are we called upon to invoke non-existent objects in a straight-faced ontology of what is going on, despite the superficially committing ways in which we think and talk.

The objects of false science, like Vulcan, differ from fictions principally in that their adherents at some point genuinely believe they exist, and go about trying to discover what they are like. The relative stability which allows people to probe and eventually falsify their supposed existence is explained in terms of intersubjective anchors in texts etc. just as for fictions. In the case of Vulcan for example, once Le Verrier had publicised his theory of a missing intramercurial planet causing observed orbital perturbations, a theory which had served spectacularly well in the case of Neptune, several astronomers claimed to have observed a new planet in transit across the face of the sun. But astronomical calculations of its supposed orbit and positions had to be inconsistent if they were to account for the discrepancies of observations of Mercury's perihelion from those predicted by Newtonian mechanics. After a prolonged period without any sightings, astronomers lost interest, though Le Verrier died still believing in Vulcan. The alternative and correct explanation of those discrepancies was only achieved when Einstein's theory of gravity superseded that of Newton.

5 Why true non-general negative existentials are true

Once upon a time, it was supposed by many philosophers that true sentences of the form 'A does not v', where 'A' is replaced by a singular or other particular term and 'v' by an intransitive verb, are true because the object or objects named by the subject term fails to exemplify the property or execute the activity whose kind is designated by the verb. Call this the subject–reject understanding of such sentences. On this understanding, if followed without exception, a sentence like

[2] Vulcan does not exist

would be true because the object denoted by the term 'Vulcan' lacks the property of existence or fails to execute the activity of existing. Now there are indeed cases where the subject⊖reject analysis accounts for the truth of a sentence, for example in the case of

    Queen Elizabeth II does not smoke

as well of ones with different kinds of predicate, for example

    The State of Liberty is not purple
    London is not north of Edinburgh

but in the Vulcan case a much better and more straightforward analysis explains the truth.

Some sentences — let us call them *atomic sentences* — are neither compounded of other sentences, nor do they contain any logical component, such as a connective, quantifier or operator. A negative sentence like [2] is not atomic, because it contains a negation. Following a tradition going back through the logical atomists Russell and Wittgenstein and a host of medievals, all the way to Aristotle, I hold that atomic sentences are, if true, true because of the existence of something. That something is a *truth-maker* for the sentence.[6] Now in the case of true non-general existential sentences of the form '*A* exists', the obvious and only plausible truth-maker is the object *A* itself. What that means is that in the analysis of such sentences the verb 'exist' serves not to pick out a property or activity like a normal predicate but only to indicate that we are dealing with this most basic kind of sentence.

That 'exist' is not a normal predicate has of course been a commonplace since Hume, but a long line of writers on philosophical logic, including Kant, Bolzano, Frege and Russell, have held that 'exists' is a second-order predicate, stating of an idea or concept that it is exemplified, and expressible in modern logic by the existential quantifier. Call this the Kantian interpretation. It is unsatisfactory on two counts. Firstly, it does not tell us what to do about singular or other non-general sentences of the form '*A* exists'. In early predicate logic, such sentences were not even considered well-formed, or if they were, it was because they were to be understood as meaning '$A = A$'. The problem there is that when only denoting terms are permitted, the sentence is trivially true, and so cannot help to account for our true negatives like [5]. Since the advent of free logic in the 1960s, this has been worked around by interpreting

    A exists

as being equivalent to

---

[6]Mulligan, Simons and Smith, 1984.

There exists x such that $x = A$

which restores the idea that existence in general is expressed by the existential quantifier, while providing a special complex predicate true just of A. It is this possibility that motivated Quine to advocate eliminating singular terms in favour of singular predicates. Granting that this equivalence holds, let us move to the second and more decisive objection. It is this. The only reason why a sentence of the form

There exists x such that Fx

is true, or why, to speak with Bolzano, the idea F is objectual, or with Frege, the concept F is non-empty, is because some individual exists which is F. To explain singular existence in terms of objectuality or non-emptiness is therefore circular, and this applies just as much to the free-logic understanding of 'A exists'. There exists something falling under the concept *being identical* to A precisely because the individual A exists. Therefore singular (more broadly, non-general) existence is the prior and more basic notion. This was indeed recognised by Brentano, for whom 'exist(s)' was no kind of predicate at all, but when used in a judgement or assertion signified the acceptance of some thing or things, while 'do(es) not exist' signified the rejection of some thing or things.

In our terms, leaving Brentano's accepting or rejecting subject out of the picture, if 'A exists' is true, it is simply because A exists, so A itself is the truth-maker for its own, personal existential sentence.

That brings us to negative existentials like [2]. The false positive opposite (contradictory) of [2] is obviously

[11] Vulcan exists

If [11] *were* true, it would be because Vulcan existed and made it true. But there is no such object as Vulcan: the subject-term of [11] and of [2] is empty. Do we need any further reason to suppose [11] false? Surely not. But, given that [2] is the contradictory of [11], we need no further reason to suppose [2] *true*. It is true *by default*, because nothing makes it false, which Vulcan would do were it to exist. But it does not exist, therefore [11] is false and [2] true, without need of anything at all to make [2] true or [11] false. The negations of false non-general existentials are true by default because nothing makes their opposites true.[7] Non-existent objects are simply not required for this purpose.

---

[7] Simons, 2008.

6 What to do about quantifiers

Given the truth of [2], it might appear admissible to infer

[12] Some things do not exist

Of course, here the quantifier 'some' could not be interpreted in the standard referential way, as

[13] There exists x such that x does not exist

for this would be simply self-contradictory. At this point, a Meinongian will leap in and say "Of course no existing things fail to exist, but I hold that we can interpret [12] consistently and truly as

[14] For some x, x does not exist

where the variable ranges over both existent and non-existent objects, and as instances verifying this truth I can offer Vulcan, Zeus, Frodo and many more." This is of course an interpretation we are striving to avoid, and will do so anon. Before proceeding however, it is worth remarking that one kind of semantics for free logic, so-called *outer domain semantics*, accepts just this picture, and indeed it is often praised by free logicians for its simplicity and intuitiveness. The domain consists of an inner one of existing individuals and an outer one of non-existing individuals, and it is among the outer domain that names like 'Vulcan' and 'Zeus' are found denotata. The standard referential (existentially loaded) quantifiers are then easily defined as

[15] There exists x such that $\ldots x \ldots$ iff for some x, x exists and $\ldots x \ldots$

and dually for the universal

[16] For all existing $x, \ldots x \ldots$ iff for all x, if x exists then $\ldots x \ldots$

The technicalities of this approach are unimpeachable, but its ontology, taken seriously, is just what we wish to avoid. Therefore *if* a way is to be found for [12] to be true, it will have to interpret the quantifier phrase 'some things' differently, and likewise for [14].

The most common way to understand the quantifiers so that they avoid ontological commitment is a substitutional interpretation. According to this, a sentence like [14] is true if and only if there exists an expression which can be substituted for 'x' so that the resulting sentence comes out true. We do in fact have such expressions to hand, such as 'Vulcan', 'Frodo' and so on, so on this interpretation we can be sure [14] is true. But is this the best way to interpret the quantifiers generally? Consider the following sentence, surely true in the actual world:

[17] For some x, no expression denotes x

Nearly all the individuals in the world are not denoted by actual expressions, but for [17] to be true on the substitutional interpretation, there would have to be an expression that denoted an object that no expression denoted, and that ain't going to happen. One way to find expressions to do the work in all such cases would be to take expressions to be necessarily existing platonic entities, with some kind of cosmic guarantee that every object be denoted by one. But we are committed to not going down the platonic route. Another ruse sometimes resorted to by logicians is to take each object as an expression denoting itself. But apart from being blatantly artificial, this collapses the distinction between objects and their names and so for existing objects amounts to the standard interpretation.

What this somewhat unsatisfactory state of affairs suggests is that we should try to combine the virtues of the standard referential interpretation with those of the substitutional interpretation. Call this the *Hybrid Interpretation*. The relevant clause in the semantics will be like this

[18] 'For some $x, \ldots x \ldots$' is true iff *either* there exists some object in the domain such that the open sentence '$\ldots x \ldots$' is true of it or there exists some expression e in the language such that the result of substituting e for 'x' in the open sentence '$\ldots x \ldots$' is true.

Obviously for a formal clause in a semantics this will need to be fleshed out with details about language, domain and interpretation, but the general point is that we are combining both accounts disjunctively. On this hybrid interpretation, both [14] and [17] can be true at once.

Now it might seem as though this hybrid view is simply *ad hoc*, concocted merely to find a way to placate both those who want to find a way for [14] to be true as well as those who are, rightly enough, sensible of the virtues of the standard interpretation. The proponents of free logic, for example, who are perfectly happy with empty singular terms when these occur unbound, are prepared to give up on the truth of [14] in order to retain the standard interpretation of the quantifiers. But here are two connected reasons why the hybrid solution is not simply *ad hoc*, and why we are not proposing two incompatible things at once. The first is that there already exists an outstanding logical system that allows [14] to be true, namely the system of Ontology of Stanisław Leśniewski. In this system, one may define a standard name 'Λ' which is necessarily empty, and then quantify to obtain [14] as a theorem, since the quantifiers are not restricted. There has never been a truly satisfactory account of how Le?niewski wanted to understand the quantifiers, and my suggestion is that the hybrid interpretation offers us the materials for giving one. But the second and arguably more important (though not unconnected) reason is that in real life, in science, and in mathematics and computing, we need a way to make sense of and regiment statements and inferences involving null cases, not just those where names lack a denotatum, but where functions are partial, predicates are undefined, and so on. A logical idiom that caters for such cases, even if it requires special clauses for such cases, is ontologically preferable to "inventing" new objects like the null set, or even worse, the null individual, and preferable also to artificial solutions regarding what happens when null cases arise, as they naturally do in real contexts, such as taking all "empty" names to denote the number 0.

Leśniewski's logic, because of its thoroughgoing extensionality, may or may not be the right way to do this, but it is a step in the right direction.

7   So then, do some things not exist?

Note that we tailored the account of hybrid semantics for a mildly regimented sentence, [14], not the ordinary English sentence [12]. There is a good reason for that. The sentence

[12] Some things do not exist

has two different possible interpretations. One is that offered by the hybrid interpretation. Here the sentence is true: we answer, "Sure, some things don't exist: Zeus, Pegasus, Vulcan, Frodo, etc. The list goes on." But because the truth of [12] then depends on the substitutional part of the hybrid interpretation, it in no way commits us to the Meinongian thesis

[19] There are some objects that do not exist

which it would do if the only way to interpret [12] as a truth were a referential interpretation with an outer domain. Avoiding this, under the standard existential interpretation, which is the other way to understand it, [12] is *false*.

The upshot is then that we can accommodate the impulses behind accepting [12] as true without being ontologically committed to non-existing objects. Speaking with a straight ontological face, i.e. interpreting the quantifier expression 'nothing' standardly, and despite there being plenty of empty terms having a sense and a rich phenomenology of meaning, we can and should affirm

[20] Nothing fails to exist.

References

[1] Bolzano, B. Theory of Science. Oxford: Oxford University Press, 2014.

[2] Husserl, E. Logical Investigations. London: Routledge, 1970.

[3] Ingarden, R. The Literary Work of Art. Evanston: Northwestern University Press, 1974.

[4] Jacquette, D. Meinong's Theory of Defective Objects. Grazer Philosophische Studien 15, 1982, 1Ð19.

[5] Jacquette, D. Brentano's Concept of Intentionality. In D. Jacquette, ed., The Cambridge Companion to Brentano. Cambridge: Cambridge University Press, 2004, 98Ð130.

[6] Mulligan, K., Simons, P. and Smith, B. Truth-Makers. Philosophy and Phenomenological Research 44 (1984), 287Ð322.

[7] Simons, P. Why the Negations of False Atomic Propositions are True. In T. de Mey and M. KeinŁnen, eds., Essays on Armstrong. Acta Philosophica Fennica 84 (2008), 15Ð36.

Closing Words

Tina Jacquette
Bern, Switzerland.

People have asked me how it was possible for one person to have such wide ranging interests as Dale had. Surely there was a mentor or succession of persons that sparked such an intellect. No doubt they are right in the way that one might say there are no new ideas but only reformulations. Dale read voraciously and widely and I cannot speak for him — it is hardly necessary given his public record. But I can attest to his fierce independence and reliance on his own internal compass for understanding, solving, and posing questions. I learned this early on when, as struggling students of philosophy, I was about to pull an all-nighter working on an epistemology essay and Dale, on giving me a goodnight kiss before his going to bed, asked me what question I was trying to answer. In realizing I didn't have a question, he said, 'well I'd start there.' So, okay, nothing revolutionary in this form of pedagogy, but it was for me at the time, because it was my first realization that posing the right question was key to establishing an argument. And argument was key to formulating, well, everything. It seems so simplistic but in the right hands it is fascinatingly complex and can lead to such interesting and unexpected results. And, over the years, I watched and learned as Dale applied his increasingly gifted hands to whatever topic he needed to understand. His drive and thirst for knowledge was boundless. And this was matched by a penchant for unconventionality, contrariness even. Freedom of exploration was paramount but even more important was arriving at conclusions independently — acceptance based on understanding (truth perhaps) not authority.

Dale had been on the debate team in high school and thought he'd study to become a lawyer until he took a philosophy course and realized that the issues and style of engagement with those issues was much more interesting than anything he had had up until then. He was hooked. He had found his niche. It was a match. His natural curiosity and desire for systematic inquiry lead to an early focus on logic and, of course, its use in argument. Dale was very good at argument. I only became aware of how much I agreed with his approach when my mother remarked to me once during a visit, 'you and Dale argue a lot.' I had to explain that it was our form of coming to a joint decision — we had been together about 15 years at that point, so I suppose we had both accepted this as the most efficient way of communicating. But it wasn't always efficient. It could be remarkably tedious and inconclusive. I remember a long car ride, probably driving from Rhode Island to Wisconsin for a holiday with extended family during which we argued for hours over whether a threat with violent consequences was worse than a perpetrated aggression of lesser violence. Neither of us convinced the other or even budged in our respective positions. But over the hours we refined our positions and honed in on what it was that was driving our divergent stances. This conversation would have driven my mother nuts. It probably would have driven most people nuts. It was the kind of geeky exchange that only appeals to people with similar interests, in this case, a love for understanding how principles are defined and conclusions derived. Luckily, we also shared a love of travel, music, film, and painting which got us outside of ourselves as well as our physical surroundings. And these interests became a major part of our lives together as Dale bored down on philosophy and I began to enjoy it from a more distant position.

Perhaps another way of gaining insight to Dale's seemingly impossible breadth is to view his personal library. When we moved to Switzerland I asked if he was really going to move all his books and if so, why. He responded, unhesitatingly, that yes he was moving his entire library because they were things he needed 'to hand.' It was only later that I understood that these books were not only valuable resources for his continuing pursuits but represented distinct moments in a life of discovery. Now, as I am engaged in cataloging these books and trying to find new owners for them, I am overwhelmed at how much they say about the man who collected and read them. Because I don't yet have the catalog completed (eventually available on LibraryThing) you'll have to take my word for how diverse and widely read he was. This isn't so unusual for scholars, it's expected really. For Dale, it wasn't the number of books, (impressive as it is), special editions, hard-bound, signed, or any of the things collectors look for that is informative. It is the sheer variety of topics, writing styles, and genres. Dale was a writer. He loved expressing himself in writing. It was his exercise, a daily necessity. He was beginning work in fiction and I was very much looking forward to that direction because his imagination was stunning and his wit laser-like. He was a force!

Our first date began a conversation that continued for over 41 years. The improbable adventures we shared reinforced a life-long love and admiration for challenge and discovery.

DALE JACQUETTE. PUBLICATIONS

COMPILED BY TINA JACQUETTE
Bern, Switzerland.

1 Books

1. *Philosophy of Mind* (Prentice Hall / Foundations of Philosophy Series, 1994) (x + 166 pp.).

2. J.N. Findlay, *Meinong's Theory of Objects and Values*, Edited with an Introduction by Dale Jacquette (Gregg Revivals) (Ashgate Publishing, 1995) (liv + 353 pp.).

3. *Meinongian Logic: The Semantics of Existence and Nonexistence* (Walter de Gruyter & Co., 1996) (xiii + 297 pp.).

4. *Schopenhauer, Philosophy, and the Arts* (edited) (Cambridge University Press, 1996) (xiii + 309 pp.); paperback edition with corrections, 2007.

5. *Wittgenstein's Thought in Transition* (Purdue University Press, 1998) (xix + 356 pp.).

6. *Six Philosophical Appetizers* (McGraw-Hill, 2001) (x + 165 pp.).

7. *Philosophical Entrées: Classic and Contemporary Readings in Philosophy*, (edited) (McGraw-Hill, 2001) (viii + 770 pp.).

8. *Symbolic Logic* (with instructor's and student solutions manuals prepared in collaboration with Andrew R. Martinez and interactive logic exercises on CD-ROM designed in collaboration with Nelson Pole) (Wadsworth Publishing, 2001) (xix + 488 pp.).

9. *David Hume's Critique of Infinity* (Brill Academic Publishers, 2001) (xvii + 384 pp.).

10. *The School of Alexius Meinong* (edited with Liliana Albertazzi and Roberto Poli) (Ashgate Publishing, 2001) (xi + 579 pp.).

11. *Ontology* (Routledge, Taylor and Francis (Acumen Imprint and McGill-Queen's University Press / Central Problems of Philosophy Series), 2002) (xv + 348 pp.).

12. *On Boole* (Wadsworth Publishing / Wadsworth Philosophers Series, 2002), (vi + 97 pp.).

13. *A Companion to Philosophical Logic* (edited) (Blackwell Publishers, 2002), (xiii + 816 pp.). (Electronic e-book NetLibrary version published to licensed distributors by Baker & Taylor, Inc.; http://www.netLibrary.com). Paperback edition with corrections 2006.

14. *Philosophy of Logic: An Anthology* (edited) (Blackwell Publishers, 2002) (xi + 372 pp.).

15. *Philosophy of Mathematics: An Anthology* (edited) (Blackwell Publishers, 2002) (xii + 428 pp.).

16. *Philosophy, Psychology, and Psychologism: Critical and Historical Readings on the Psycho-logical Turn in Philosophy* (edited) (Kluwer Academic Publishing / Philosophical Studies Series, volume 91, 2003) (xiii + 339 pp.); paperback edition, 2010. Published by Kluwer in electronic e-book format, 2004.

17. *Pathways in Philosophy: An Introductory Guide with Readings* (Oxford University Press, 2004) (xv + 555 pp.).

18. *The Cambridge Companion to Brentano* (edited) (Cambridge University Press, 2004) (xxii + 322 pp.). (Electronic e-book NetLibrary version published to licensed distributors by Cambridge University Press; http://www.ebookmall.com/ebook/207919-ebook.htm).

19. *The Philosophy of Schopenhauer* (Routledge, Taylor and Francis (Acumen Imprint, McGill-Queen's University Press), 2005) (xiv + 305 pp.)

20. *Philosophy of Logic* (edited) Handbook of the Philosophy of Science series, edited by Dov Gabbay, John Woods and Paul Thagard (North-Holland Press (Elsevier), 2007) (xvi + 1202 pp.) (forthcoming also in Russian and Chinese translations).

21. *Journalistic Ethics: Moral Responsibility in the Media* (Routledge, Taylor and Francis (Prentice Hall Imprint), 2007) (xv + 300 pp.).

22. *Gottlob Frege, The Foundations of Arithmetic: A Logical-Mathematical Investigation into the Concept of Number*, Translation and Introduction With Critical Commentary by Dale Jacquette (of Die Grundlagen der Arithmetik: eine logisch mathematische Untersuchungen über den Begriff der Zahl, 1884) (Routledge, Taylor and Francis (Longman Imprint) 2007) (xxxii + 112 pp.).

23. *After 'On Denoting': Themes from Russell and Meinong* (edited with Nicholas Griffin and Kenneth Blackwell) (Hamilton: McMaster University for The Bertrand Russell Research Centre, 2007) (contents also published as special issue of Russell: The Journal of Bertrand Russell Studies, 2007, below) (188 pp).

24. *Russell versus Meinong: The Legacy of 'On Denoting'* (edited with Nicholas Griffin), Routledge, 2009) (xiii + 384 pp.); paperback edition 2011.

25. *Dialogues on the Ethics of Capital Punishment* (Rowman & Littlefield, 2009) (vii + 139 pp.).

26. *Reason, Method, and Value: A Reader on the Philosophy of Nicholas Rescher* (edited and with a critical introduction) (Ontos Verlag, 2009) (vi + 643 pp.).

27. *Philosophy of Mind: The Metaphysics of Consciousness* (Continuum Books, 2009) (xiii + 307 pp.) (2nd revised edition of Philosophy of Mind (1994)).

28. *Logic and How it Gets That Way* (Routledge, Taylor and Francis (Acumen Imprint), 2010) (xiv + 306 pp.); paperback edition 2013.

29. *Cannabis: What Were We Just Talking About?* (edited) (Wiley-Blackwell, 2010) (xix + 241 pp.). La philosophie et le cannabis (edited) (French translation of Wiley- Blackwell 2010) (Original Books, 2011) (261 pp.). Cannabis: 'De qué estamos hablando?' (edited) (Spanish translation of Wiley-Blackwell 2010) (Canamo Ediciones, 2013) (248 pp.).

30. *Possible Worlds: Logic, Semantics and Ontology* (edited with Guido Imaguire) (Philosophia Verlag, 2010) (287 pp.).

31. *Around and Beyond the Square of Opposition* (edited with Jean-Yves Beziau) (New York: Birkhauser, Springer Verlag, 2012) (x + 379 pp.).

32. *George Berkeley, Three Dialogues Between Hylas and Philonous*, edited by Dale Jacquette (Toronto: Broadview Editions, 2013) (256 pp.).

33. *Alexius Meinong: The Shepherd of Non-Being* (Cham, Heidelberg, New York, Dordrecht, London: Springer Verlag, 2015), Synthese Library 360 (xxxii + 434 pp).

34. *Schopenhauer on the Metaphysics of the Unconscious*, Zürcher Philosophische Vorträge 7, herausgegeben von Wolfgang Rother (Basel: Schwabe Verlag, 2015) (32 pp.).

35. *The Bloomsbury Companion to the Philosophy of Consciousness* (edited) (Bloomsbury Academic, 2018), (504 pp).

36. *Frege: A Philosophical Biography* (Cambridge University Press, announced publication March 2018).

37. *David Hume's Experimental Philosophy of Religion* (Palgrave Macmillan (not completed)).

38. *Skeptical Studies in Analytic Philosophy of Religion* (Springer Verlag, Boston Studies in Philosophy of Religion series (not completed)).

39. *From Epistemology to Philosophy of Science: Reflections on Justifying Scientific Knowledge* (Springer Verlag, Philosophical Studies series (not completed)).

2 Series, Journal and Journal Guest Editing

1. Editor, American Philosophical Quarterly, 2002-2005.

2. Co-Editor (with Hans Burkhardt, Guido Imaguire, Ludger Jansen, und Maarten Hoenen), and Review Editor, *Ontologica: Journal of Metaphysics, Ontology, and Applied Ontology*.

3. Series General Editor, *New Dialogues in Philosophy*, eight commissioned titles for Rowman & Littlefield on philosophical topics in logic and metaphysics, epistemology, philosophy of science, and value theory. 2006-2016.

4. Series Co-Editor-in-Chief (with Wolfgang L. Gombocz), *Resources*, Philosophia Verlag. 2008+.

5. *The Journal of Value Inquiry*, 35, 2001, special issue on 'Aristotle's Theory of Value', guest edited by Dale Jacquette.

6. Philosophy and Rhetoric, 30, 1997, special issue on 'The Dialectics of Psychologism', guest edited by Dale Jacquette.

7. Russell: Journal of the Bertrand Russell Society, special issue on the 100th anniversary of 'After 'On Denoting'': Themes from Russell and Meinong', guest co-edited by Nicholas Griffin, Dale Jacquette, and Kenneth Blackwell, 27, 2007.

8. Topoi: An International Review of Philosophy, special issue on 'Logic, Meaning, and Truth-Making States of Affairs in Philosophical Semantics', guest edited by Dale Jacquette, 2010.

9. The Monist: An International Quarterly Journal of General Philosophical Inquiry, special issue on 'Formal and Intentional Semantics', guest edited by Dale Jacquette, 96, 2013.

3 Articles in Refereed Journals

1. "Thoughts on Twin Earth", *Metaphysica: International Journal for Ontology & Metaphysics*, 18 (1), 2017, 33-59.

2. "Brentano's Signature Contributions to Scientific Philosophy", *Brentano Studien*, 14, 2016, 127-156.

3. "Subalternation and Existence Presuppositions in an Unconventionally Formalized Canonical Square of Opposition", *Logica Universalis*, 10, 2016, 191-213. DOI 10.1007/s11787-016-0147-y.

4. "Marx and Industrial Age Aesthetics of Alienation", *Cultura: International Journal of Philosophy of Culture and Axiology*, 13, 2016, 89-105.

5. "Contemporary and Future Directions of Analytic Philosophy: Commentary on Jaakko Hintikka, 'Philosophical Research: Problems and Proposals'", *Diogenes* 2016, DOI 10.1177/039219216640718; Chinese translation, Diogenes, 62, 2015, 50-63.

6. "Condillac's Analytic Dilemma", *History of Philosophy Quarterly*, 32, 2015, 141-160.

7. "Categories and Preferences Among Category Systems", *The Monist*, 98, 2015, 268-289. DOI 10.1093/monist/onv010.

8. "Salinger's World of Adolescent Disillusion", *Philosophy and Literature*, 39, 2015, 156- 177.

9. "Later Wittgenstein on the Invention of Games", *Linguistic and Philosophical Investigations*, 14, 2015, 19-38.

10. "Slingshot Arguments and the Intensionality of Identity", *European Journal of Analytic Philosophy*, 11, 1, 2015, 5-21.

11. "Berkeley's Unseen Horse and Coach", *Idealistic Studies*, 45, 3, 2015, 247-264.

12. "Semantics and Pragmatics of Referentially Transparent and Referentially Opaque Belief Ascription Sentences", *Philosophia: Philosophical Quarterly of Israel*, 43, 2015, Online First DOI 10.1007/s11406-015-9604-8.

13. "Computable Diagonalizations and Turing's Cardinality Paradox", *Zeitschrift für allgemeine Wissenschaftstheorie* (Journal for General Philosophy of Science), 45, 2014, 239-262. Online First Publication 2014: DOI 10.1007/s10838-014-9244-x.

14. "Newton's Metaphysics of Space as God's Emanative Effect", *Physics in Perspective*, 16, 344-370. Online First Publication 2014: DOI 10.1007/s00016-014-0142-8.

15. "Later Wittgenstein's Anti-Philosophical Therapy", *Philosophy*, 89, 2014, 251-272. DOI 10.1017/S0031819114000011.

16. "Intentionality in Reference and Action", *Topoi: An International Review of Philosophy*, 33, 2014, 255-262. DOI 10.1007/s11245-013-9165-z.

17. "Toward a Neoaristotelian Inherence Philosophy of Mathematical Entities', *Studia Neoaristotelica*, 11, 2014, 159-204.

18. "Kripke on the Necessity of Identity", *Analysis and Metaphysics*, 13, 2014, 7-26.

19. "Socrates on the Moral Mischief of Misology", *Argumentation*, 28, 2014, 1-17. DOI 10.1007/s10503-013-9298-7.

20. "Against Logically Possible World-Relativized Existence", Metaphysica: International *Journal for Ontology & Metaphysics*, 15, 2014, 85-98. DOI 10.1515/mp-2014-0006.

21. "Collingwood on Religious Atonement", *International Journal for Philosophy of Religion*, 76, 2014, 151-170. Online First Publication 2014: DOI 10.1007/s11153- 014-9456-3.

22. "Art, Expression, Perception, and Intentionality", *Journal of Aesthetics and Phenomenology*, 1, 2014, 63-90.

23. "Beccaria on Discounting Intentions in Adjudicating Punishments for Crimes", *Philosophical Inquiries*, 2, 2014, 107-120.

24. "Algebra of Classical Theoretical Term Reductions in the Sciences", *Symposion: Theoretical and Applied Inquiries in Philosophy and Social Sciences*, 1, 2014, 51-67.

25. "Collective Referential Intentionality in the Semantics of Dialogue", *Studies in Logic, Grammar and Rhetoric*, special issue on Pragmatics and Dialectics of Argument, edited by Katarzyna Budzynska, Frans van Eemeren, and Marcin Koszowy, 36 (49), 2014, 143-159. Preprint On-line DOI 10.2478/slgr-2014-0007.

26. "Searle on Collective Intentionality and the Foundations of Symbolic Social Institutional Status", *Balkan Journal of Philosophy*, 6, 2014, 21-32 [Invited essay for special issue on Collective Intentionality].

27. "Syntactical Constraints on Definitions", *Metaphilosophy*, 44, 2013, 145-156.

28. "About Nothing", *Humana-Mente: Journal of Philosophical Studies*, special issue, 'Meinong Strikes Again: Meinong's Ontology in the Current Philosophical Debate", 2013, 95-118.

29. "Phenomenological Thought Content, Intentionality, and Reference in Putnam's Twin- Earth", *The Philosophical Forum*, 44, 2013, 69-87.

30. "Violence as Intentionally Inflicting Forceful Harm", *Revue Internationale de Philosophie*, 67, 2013, 293-322.

31. "Qualities, Relations, and Property Exemplification", *Axiomathes*, special issue on Exemplification: Between the Particular and the Universal, edited by Javier Cumpa Arteseros, 23, 2013, 381-399.

32. "Tendances de la philosophie analytique", Diogenes, 242, 2013, 50-63.

33. "Faith as a Mustard Seed", *Forum Philosophicum: International Journal for Philosophy*, 17, 2012, 141-173.

34. "Causal Relevance and Relevant Causation", *Journal of Logic and Computation*, 22, 2012, 101-112.

35. "Hume on Color Knowledge, With an Application to Jackson's Thought Experiment", *History of Philosophy Quarterly*, 29, 2012, 355-371.

36. "Anselm's Metaphysics of Nonbeing", *European Journal for Philosophy of Religion*, 4, 2012, 27-48.

37. "Justification and Truth Conditions in the Analysis of Knowledge", *Logos & Episteme*, 3, 2012, 429-447.

38. "Intuition in Some Methods of Philosophical Inquiry", Al-Mukhatabat: Philosophical *Journal for Logic, Epistemology and Analytic Philosophy* (Tunesia) (electronic format), 1, 2012, 33-51.

39. "A Dialogue on Metaphysics", Philosophy Now, 92, 2012, 33.

40. "Some Monkey Devours Every Raisin", *Journal of Applied Non-Classical Logics*, 21, 2011, 201-209.

41. "Modal Objection to Naive Leibnizian Identity", *History and Philosophy of Logic*, 32, 2011, 107-118.

42. "Intentionality as a Conceptually Primitive Relation", *Acta Analytica*, 26, 2011, 15-35.

43. "Frege on Identity as a Relation of Names", *Metaphysica: International Journal for Ontology & Metaphysics*, 12, 2011, 51-72.

44. "Brentano on Aristotle's Categories: First Philosophy and the Manifold Senses of Being", *Revue roumaine de Philosophie*, 55, 2011, 169-197.

45. "A Note on Epistemic Naivety in Marx and Engels", *The Critical Review*, 23, 2011, 117- 122.

46. "How (Not) to Justify Induction", *Kriterion: Journal of Philosophy*, 24, 2011, 1-18.

47. "Enhancing the Diagramming Method in Informal Logic", *Argument*, 1, 2011, 327-360.

48. "Axiom of Infinity and Plato's Third Man", *Russell: The Journal of Bertrand Russell Studies*, 30, 2010, 5-13.

49. "Circularity or Lacunae in Tarski's Truth-Schemata", *Journal of Logic, Language and Information*, 19, 2010, 315-326.

50. "Truth Breakers", *Topoi: An International Review of Philosophy*, special issue on 'Logic, Meaning, and Truth-Making States of Affairs in Philosophical Semantics', guest edited by Dale Jacquette, 29, 2010, 153-163; 'Introduction', 87-89.

51. "Liar Paradox and Substitution into Intensional Contexts", *Polish Journal of Philosophy*, 4, 2010, 119-147.

52. "Self-Justificatory versus Extra-Justificatory Foundations of Knowledge", *Epistemologia: An International Journal for the Philosophy of Science*, 33, 2010, 111-134.

53. "Metaphysics of Meinongian Aesthetic Value", *Meinong Studies*, 4, 2010, special issue on The Aesthetics of the Graz School, edited by Venanzio Raspa, 53-86.

54. "Logic of Courage and Wisdom in Plato's Protagoras 349e-351b", *Logical Analysis and History of Philosophy*, 12, 2009, 48-69.

55. "Collingwood on Historical Authority and Historical Imagination", *Journal of the Philoso-phy of History*, 3, 2009, 55-78.

56. "Revisionary Early-Peircean Predicate Logic Without Proper Names", *Transactions of the Charles S. Peirce Society*, 45, 2009, 177-213.

57. "Deductivism in Formal and Informal Logic", *Studies in Logic, Grammar and Rhetoric*, special issue on Informal Logic and Argumentation Theory, guest edited by Marcin Koszowy, 16 (29), 2009, 189-216.

58. "Object Theory Logic and Mathematics: Two Essays by Ernst Mally (Translation and Critical Commentary)", *History and Philosophy of Logic*, 29, 2008, 167-182.

59. "Mathematical Discovery and Proof Reductio ad Absurdum", *Informal Logic: Reason and Argumentation in Theory and Practice*, 28, 2008, 242-261.

60. "Denying the Liar Reaffirmed", *Polish Journal of Philosophy*, 2, 2008, 143-157.

61. "Logic of the Preface Paradox", *Principia: International Journal of Epistemology*, 12, 2008, 203-216.

62. "Innate Psychologism in Sigwart's Foundations of Mathematics", *Erwägen - Wissen - Ethik*, 19, 2008, special issue on Christoph Sigwart's Logik, Die

Zahlbegriffe, 571- 572; also, "Reading Kant Between the Lines of Sigwart's Philosophy of Arithmetic", 588-589; "Philosophical Afterthoughts", 602-603.

63. "Preface: After Russell, After Meinong", *Russell: Journal of the Bertrand Russell Society*, special issue on the 100th anniversary of 'On Denoting', guest co-edited with Nicholas Griffin and Kenneth Blackwell, 27, 2007, 5-6.

64. "Burleigh's Paradox", *Philosophy*, 82, 2007, 437-448.

65. "Denying the Liar", *Polish Journal of Philosophy*, 1, 2007, 91-98.

66. "Schopenhauer's Proof That Thing-in-Itself is Will", *Kantian Review*, 12, 2007, 76-108.

67. "Two Sides of Any Issue", *Argumentation*, 21, 2007, 115-127.

68. "Deductivism and the Informal Fallacies", *Argumentation*, 21, 2007, 335-347.

69. "Essential Concepts of Systematic Logic" (Chinese translation), *Journal of Peking University* (Philosophy and Social Sciences Edition), *Logic Forum*, 44, 2007, 35-42.

70. "Manometrul lui Wittgenstein si argumentul limgajului privat" (Romanian translation of "Wittgenstein's Manometer and the Private Language Argument", 1998, below), *Linguistic and Philosophical Investigations*, 6, 2007, 120-148.

71. "Bochenski on Property Identity and the Refutation of Universals", *Journal of Philosophical Logic*, 35, 2006, 293-316.

72. "Propositions, Sets, and Worlds", *Studia Logica: An International Journal for Symbolic Logic*, special double issue 'Ways of Worlds' on 40 Years of Possible Worlds Semantics, Vol. 1: On Possible Worlds and Related Notions, edited by Vincent F. Hendricks and Stig Andur Pedersen, 82, 2006, 337-343.

73. "Intention, Meaning, and Substance in the Phenomenology of Abstract Painting", *The British Journal of Aesthetics*, 46, 2006, 38-58.

74. "Supervenience of Qualia and Intentionality", *Philo: A Journal of Philosophy*, 9, 2006, 145-164.

75. "An Elementary Deductive Logic Exercise: Maximus Tyrius's Proof That There is No Injustice", *Teaching Philosophy, Logic Notes*, 29, 2006, 45-52.

76. "Collingwood Against Metaphysical Realism", *Collingwood and British Idealism Studies*, 12, 2006, 103-114.

77. "Applied Mathematics in the Sciences", *The Croatian Journal of Philosophy*, 6, 2006, 237-267.

78. "Adevar si fictiune in critica lui David Lewis asupra semanticii meinongiene" (Romanian translation of "Truth and Fiction in David Lewis's Critique of Meinongian Semantics", 2001, below), *Linguistic and Philosophical Investigations*, 5, No. 1, 2006, 55-86.

79. "Antireductionsimul lui Searle" (Romanian translation of "Searle's Antireductionism", 2002, below), *Linguistic and Philosophical Investigations*, 5, No. 2, 2006, 95-119.

80. "Margolis on History and Nature", *Metaphilosophy*, 36, 2005, 568-577.

81. "Nonstandard Modal Semantics and the Concept of a Logically Possible World", *Philosophia Scientiae*, 9, 2005, special issue on 'Aperçus philosophiques en logique et en mathématiques', edited by Gerhard Heinzmann and Manuel Rebuschi, 239-258.

82. "Indeterminarea Cuantica i Argumentul Wittgensteinian al Limbajului Privat" (Romanian translation of "Quantum Indeterminacy and Wittgenstein's Private Language Argument", 1999, below), *Analysis and Metaphysics*, 4, 2005, 125-147.

83. "Grelling's Revenge", *Analysis*, 64, 2004, 251-256.

84. "Idealism and Williams's Semantic Paradox", *Philosophical Investigations*, 27, 2004, 117-128.

85. "Theory and Observation in the Philosophy of Science", *Logical Analysis and History of Philosophy*, 7, special issue on History of the Philosophy of Nature ('Geschichte der Naturphilosophie'), 7, 2004, 177-196.

86. "The Soundness Paradox", *Logic Journal of the Interest Group in Pure and Applied Logics* (IGPL), 11, 2003, 547-556.

87. "Thomas Reid on Natural Signs, Natural Principles, and the Existence of the External World", *The Review of Metaphysics*, 57, 2003, 279-300.

88. "Probability, Practical Reasoning, and Conditional Statements of Intent", *The Harvard Review of Philosophy*, 11, 2003, 101-113.

89. "Conditional Intent in the Strange Case of Murder by Logic", *Logic and Logical Philosophy*, 12, 2003, 301-316.

90. "Meinong on the Phenomenology of Assumption", *Studia Phænomenologica*, special issue on 'The School of Brentano and Husserlian Phenomenology', 3, 2003, 155-177.

91. "Plato on the Parts of the Soul", *Epoché: A Journal for the History of Philosophy*, 8, 2003, 43-68.

92. "Socrates on Persuasion, Truth, and Courtroom Argumentation", *Inquiry: Critical Thinking Across the Disciplines*, 22, 2003, 33-41.

93. "Searle's Antireductionism", *Facta Philosophica*, 4, 2002, 143-166.

94. "Hume on Infinite Divisibility and the Negative Idea of a Vacuum", *The British Journal for the History of Philosophy*, 10, 2002, 413-435.

95. "Brentano's Scientific Revolution in Philosophy", Spindel Conference 2001, Origins: The Common Sources of Analytic and Phenomenological Traditions, *Southern Journal of Philosophy*, Spindel Conference Supplement, 40, 2002, 193-221.

96. "Analysis of Quantifiers in Wittgenstein's Tractatus: A Critical Survey", *Logical Analysis and History of Philosophy*, 4, 2001, 191-202. 97. "Of Time and the River in Kant's Refutation of Idealism", *History of Philosophy Quarterly*, 18, 2001, 297-310.

97. "Psychologism Revisited in Logic, Metaphysics, and Epistemology", *Metaphilosophy*, 32, 2001, 261-278.

98. "Truth and Fiction in David Lewis's Critique of Meinongian Semantics", *Metaphysica: International Journal for Ontology and Metaphysics*, 2, 2001, 73-106.

99. "Introduction", *The Journal of Value Inquiry*, special issue on 'Aristotle's Theory of Value', guest edited by Dale Jacquette, 35, 2001, 303-308.

100. "Aristotle on the Value of Friendship as a Motivation for Morality", *The Journal of Value Inquiry*, special issue on 'Aristotle's Theory of Value', guest edited by Dale Jacquette, 35, 2001, 371-389.

101. "Two Kinds of Potentiality: A Critique of McGinn on the Ethics of Abortion", *Journal of Applied Philosophy*, 17, 2001, 79-86.

102. "Fin de Siècle Austrian Thought and the Rise of Scientific Philosophy", *History of European Ideas*, 27, 2001, 307-315.

103. "Carl Stumpf on the Ontology of Relations", *Brentano Studien*, 9, 2000-2001, 229-250.

104. "Identity, Intensionality, and Moore's Paradox", *Synthese*, 123, 2000, 279-292.

105. "Conundrums of Conditionals in Contraposition", *Nordic Journal of Philosophical Logic*, 4, 2000, 117-126.

106. "Confessions of a Meinongian Logician", *Grazer Philosophische Studien*, 58-59, 2000, 151-180.

107. "Goodman on the Concept of Style", *The British Journal of Aesthetics*, 40, 2000, 452-466.

108. "Liar Paradox and Metaparadox", *Sats: Nordisk Tidsskrift for Filosofi*, 1, 2000, 93-104.

109. "An Internal Determinacy Metatheorem for Lukasiewicz's Aussagenkalküls", *Bulletin of the Section of Logic*, 29, 2000, 115-124.

110. "Schopenhauer on the Ethics of Suicide", *Continental Philosophy Review*, 33, 2000, 43-58.

111. "Demonstratives and the Logic of the Self", *Philosophical Papers*, 28, 1999, 1-23.

112. "Quantum Indeterminacy and Wittgenstein's Private Language Argument", *Philosophical Explorations: An International Journal for the Philosophy of Mind and Action*, 2, 1999, 79-95.

113. "Wittgenstein's Anthropologism in Logic, Philosophy, and the Social Sciences", *Social Epistemology*, 13, 1999, 303-322.

114. "L'Antropologismo Wittgensteiniano nella Logica, nella Filosofia e nelle Scienze Sociali", *Studi Perugini* (Italian translation of "Wittgenstein's Anthropologism in Logic, Philosophy, and the Social Sciences", 1999, above), 7, 1999, 159-189.

115. "Intentionality on the Instalment Plan", *Philosophy*, 73, 1998, 63-79.

116. "Paraconsistent Logical Consequence", *Journal of Applied Non-Classical Logics*, 8, 1998, 337-351.

117. "Wittgenstein's Manometer and the Private Language Argument", *History of Philosophy Quarterly*, 15, 1998, 99-126.

118. "Aristotle's Refutation of the Universal Good", *The Journal of Value Inquiry*, 32, 1998, 301-324.

119. "The Devil's Dilemma in Flaubert's Saint Antony", *The Heythrop Journal: A Quarterly Review of Philosophy and Theology*, 39, 1998, 140-147.

120. "La NeurobiologŠa y el Alma", *Estudios Filoso' ficos* (Spanish translation), 47, 1998, 143-156.

121. "Wittgenstein on the Transcendence of Ethics", *Australasian Journal of Philosophy*, 75, 1997, 304-324.

122. "Conceivability, Intensionality, and the Logic of Anselm's Modal Argument for the Existence of God", *International Journal for Philosophy of Religion*, 42, 1997, 163- 173.

123. "The Dialectics of Psychologism", *Philosophy and Rhetoric*, special issue on 'The Dialectics of Psychologism', guest edited by Dale Jacquette, 30, 1997, v-viii.

124. "Psychologism the Philosophical Shibboleth", *Philosophy and Rhetoric*, special issue on 'The Dialectics of Psychologism', guest edited by Dale Jacquette, 30, 1997, 312-331.

125. "Reflections on Mally's Heresy", *Axiomathes*, 8, 1997, 163-180.

126. "Helmholtz's Conservation of Energy Proof for Mind-Body Identity", *Philosophy in Science*, 7, 1997, 129-136.

127. "The Validity Paradox in Modal S5", *Synthese*, 109, 1996, 47-62.

128. "Is Nondefectively Justified True Belief Knowledge?", *Ratio*, 9, 1996, 115-127.

129. "Lloyd on Intrinsic Natural Representation in Simple Mechanical Minds", *Minds and Machines*, 6, 1996, 47-60.

130. "Hume on Infinite Divisibility and Sensible Extensionless Indivisibles", *Journal of the History of Philosophy*, 34, 1996, 61-78.

131. "What Would a Cerebroscope Do?", *Journal of the British Society for Phenomenology*, 27, 1996, 188-199.

132. "Kant on Unconditional Submission to the Suzerain", *History of Philosophy Quarterly*, 13, 1996, 117-131.

133. "Many-Valued Deontic Predications", *Logique et Analyse*, 155-156, 1996, 243-254.

134. "Charity and the Reiteration Problem for Enthymemes", *Informal Logic*, 18, 1996, 1-15.

135. "Socrates' Ironic Image of Meno", *The Personalist Forum*, 12, 1996, 123-134.

136. "Adversus Adversus Regressum (Against Infinite Regress Objections)", *The Journal of Speculative Philosophy*, 10, 1996, 105-119.

137. "On Defoliating Meinong's Jungle", *Axiomathes*, special issue on 'The Philosophy of Alexius Meinong', 7, 1996, 17-42. Electronic version available from SpringerLink at http://dx.doi.org/10.1007/BF02357196.

138. "Descartes' Lumen naturale and the Cartesian Circle", *Philosophy & Theology*, 9, 1996, 273-320.

139. "Untitled", *Philosophy and Literature*, 19, 1995, 102-105.

140. "Hume's Aesthetic Psychology of Distance, Greatness and the Sublime", *The British Journal for the History of Philosophy*, 3, 1995, 89-112.

141. "Color and Armstrong's Color Realism Under the Microscope", *Studies in History and Philosophy of Science*, 26, 1995, 389-406.

142. "Meinong's Concept of Implexive Being and Nonbeing", *Grazer Philosophische Studien*, 50, 1995, 233-271.

143. "The Blue Banana Trick: Dennett on Jackson's Color Scientist", *Theoria*, 61, 1995, 217- 230.

144. "Zeno of Citium on the Divinity of the Cosmos", *Studies in Religion / Sciences Religieuses*, 24, 1995, 415-431.

145. "Object Theory Foundations for Intensional Logic", *Acta Analytica*, 13, 1995, 33-63.

146. "Virtual Relations", *Idealistic Studies*, 25, 1995, 141-154 (see below).

147. "Virtual Relations vs. Virtual Universals: Essay, Comments, and Reply" (with William J. Rapaport), Center for Cognitive Science Technical Report 95-10 (Buffalo: State University of New York Technical Report Series, 1995), 1-20.

148. "Obligations Under Causal Constraints", *Synthese*, 99, 1994, 307-310.

149. "Formalization in Philosophical Logic", *The Monist*, 77, 1994, 358-375.

150. "Infinite Divisibility in Hume's First Enquiry", *Hume Studies*, 20, 1994, 219-240.

151. "Tarski's Quantificational Semantics and Meinongian Object Theory Domains", *Pacific Philosophical Quarterly*, 75, 1994, 88-107.

152. "Knowledge, Skepticism, and the Diallelus", *International Philosophical Quarterly*, 34, 1994, 191-198.

153. "The Type-Token Distinction in Margolis's Aesthetics", *The Journal of Aesthetics and Art Criticism*, 52, 1994, 299-307.

154. "On the Designated Student and Related Induction Paradoxes", *Canadian Journal of Philosophy*, 24, 1994, 583-592.

155. "A Meinongian Theory of Definite Description", *Axiomathes*, 5, 1994, 345-359.

156. "Meinongian Logic and Anselm's Ontological Proof for the Existence of God", *The Philosophical Forum*, 25, 1994, 231-240.

157. "Many Questions Begs the Question (But Questions Do Not Beg the Question)", *Argumentation*, 8, 1994, 283-289.

158. "Wittgenstein on Private Language and Private Mental Objects", *Wittgenstein Studien*, 1, 1994, Article 12 (computer disk format textname: 12-1-94.TXT) (89K) (c. 29 pp.).

159. "Intentionality and the Myth of Pure Syntax", *Protosoziologie*, 6, 1994, 76-89; 331-333.

160. "Schopenhauer on the Antipathy of Aesthetic Genius and the Charming", *History of European Ideas*, 18, 1994, 373-385.

161. "A Turing Test Conversation", *Philosophy*, 68, 1993, 231-233.

162. "Who's Afraid of the Turing Test?", *Behavior and Philosophy*, 20-21, 1993, 63-74.

163. "Logical Dimensions of Question-Begging Argument", *American Philosophical Quarterly*, 30, 1993, 317-327.

164. "Kant's Second Antinomy and Hume's Theory of Extensionless Indivisibles", *Kant- Studien*, 84, 1993, 38-50.

165. "Pollock on Token Physicalism, Agent Materialism, and Strong Artificial Intelligence", *International Studies in the Philosophy of Science*, 7, 1993, 127-140.

166. "Chisholm on Persons as Entia Successiva and the Brain-Microparticle Hypothesis", *The Modern Schoolman: A Quarterly Journal of Philosophy*, 70, 1993, 99-113.

167. "Reconciling Berkeley's Microscopes in God's Infinite Mind", *Religious Studies*, 29, 1993, 453-463.

168. "A Dialogue on Zeno's Paradox of Achilles and the Tortoise", *Argumentation*, 7, 1993, 273-290.

169. "Metaphilosophy in Wittgenstein's City", *International Studies in Philosophy*, 25, 1993, 27-35.

170. "Wittgenstein's Critique of Propositional Attitude and Russell's Theory of Judgment", *Brentano Studien*, *4*, 1992-1993, 193-220.

171. "Schopenhauer's Circle and the Principle of Sufficient Reason", *Metaphilosophy*, 23, 1992, 279-287.

172. "Contradiction", Philosophy and Rhetoric, 25, 1992, 365-390.

173. "A Deflationary Resolution of the Surprise Event Paradox", *Iyyun: The Jerusalem]Philosophical Quarterly*, 41, 1992, 335-349.

174. "Buridan's Bridge", *Philosophy*, 66, 1991, 455-471.

175. "Moral Dilemmas, Disjunctive Obligations, and Kant's Principle that 'Ought' Implies 'Can'", *Synthese*, 88, 1991, 43-55.

176. "On the Completeness of a Certain System of Arithmetic of Whole Numbers in which Addition Occurs as the Only Operation", translation of and commentary on Mojzesz Presburger, "Über die Vollständigkeit eines gewissen Systems

der Arithmetik ganzer Zahlen, in welchem die Addition als einzige Operation hervortritt", *History and Philosophy of Logic*, 12, 1991, 225-233.

177. "Democracy and the Perils of Informed Consent", *Contemporary Philosophy*, 13, 1991, 13-18.

178. "The Origins of Gegenstandstheorie: Immanent and Transcendent Intentional Objects in Brentano, Twardowski, and Meinong", *Brentano Studien*, 3, 1990-1991, 277-302.

179. "Fear and Loathing (And Other Intentional States) in Searle's Chinese Room", *Philosophical Psychology*, 3, 1990, 287-304.

180. . "Intentionality and Stich's Theory of Brain Sentence Syntax", *The Philosophical Quarterly*, 40, 1990, 169-182.

181. "Wittgenstein and the Color Incompatibility Problem", *History of Philosophy Quarterly*, 7, 1990, 353-365.

182. "A Fregean Solution to the Paradox of Analysis", *Grazer Philosophische Studien*, 37, 1990, 59-73.

183. "Aesthetics and Natural Law in Newton's Methodology", *Journal of the History of Ideas*, 51, 1990, 659-666.

184. "The Sophist's Dilemma in Plato's Meno", *Cogito*, 4, 1990, 112-119.

185. "Borges' Proof for the Existence of God", *The Journal of Speculative Philosophy*, 4, 1990, 83-88.

186. "Searle's Intentionality Thesis", *Synthese*, 80, 1989, 267-275.

187. "Adventures in the Chinese Room", *Philosophy and Phenomenological Research*, 49, 1989, 605-623.

188. "Stich Against De Dicto - De Re Ambiguity", *Philosophical Psychology*, 2, 1989, 223- 230.

189. "Mally's Heresy and the Logic of Meinong's Object Theory", *History and Philosophy of Logic*, 10, 1989, 1-14.

190. "Intentional Semantics and the Logic of Fiction", *The British Journal of Aesthetics*, 29, 1989, 168-176.

191. "The Hidden Logic of Slippery Slope Arguments", *Philosophy and Rhetoric*, 22, 1989, 59-70.

192. "Dualities of Self-Non-Application and Infinite Regress" (with Henry W. Johnstone, Jr.), *Logique et Analyse*, 125-126, 1989, 29-40.

193. "Modal Meinongian Logic", *Logique et Analyse*, 125-126, 1989, 113-130.

194. "On the Objects' Independence from Thought", translation of and commentary on Ernst Mally, "Über die Unabhängigkeit der Gegenstände vom Denken", *Man and World*, 22, 1989, 215-231.

195. "Presupposition and Foundational Asymmetry in Metaphysics and Logic", *Philosophia Mathematica: International Journal for Philosophy of Mathematics*, 4, 1989, 15-22.

196. "Epistemic Blood from Logical Turnips", *Philosophy and Rhetoric*, 22, 1989, 203-211.

197. "Knowledge and Aesthetic Appreciation", *Art & Academe*, 2, 1989, 29-41.

198. "Explanatory Limitations of Sociobiology", *Journal of Social Philosophy*, 19, 1988, 56- 62.

199. "Metamathematical Criteria for Minds and Machines", *Erkenntnis*, 27, 1987, 1-16.

200. "Intentionality and Intentional Connections", *Philosophia: Philosophical Quarterly of Israel*, 17, 1987, 13-31.

201. "Kripke and the Mind-Body Problem", *Dialectica*, 41, 1987, 293-300.

202. "Twardowski on Content and Object", *Conceptus: Zeitschrift für Philosophie, Österreichische Philosophen und ihr Einfluss auf die Analytische Philosophie der Gegenwart*, 21, 1987, Band 2, 193-199.

203. "Intentionality and Intensionality: Quotation Contexts and the Modal Wedge", *The Monist*, 59, 1986, 598-608.

204. "Margolis on Emergence and Embodiment", *The Journal of Aesthetics and Art Criticism*, 44, 1986, 257-261.

205. "Émergence et incorporation selon Margolis", *Philosophiques*, 13, 1986, 53-63 (French translation).

206. "Forrester's Paradox", *Dialogue*, 25, 1986, 761-763.

207. "The Uniqueness Problem in Kant's Transcendental Doctrine of Method", *Man and World*, 19, 1986, 425-438.

208. "Meinong's Doctrine of the Modal Moment", *Grazer Philosophische Studien*, 25-26, 1985-1986, 423-438.

209. "Sensation and Intentionality", *Philosophical Studies*, 47, 1985, 429-440.

210. "Logical Behaviorism and the Simulation of Mental Episodes", *The Journal of Mind and Behavior*, 6, 1985, 325-332.

211. "Wittgenstein on Frege's Urteilstrich", *International Logic Review*, 32, 1985, 79-82.

212. "Berkeley's Continuity Argument for the Existence of God", *The Journal of Religion*, 65, 1985, 1-14.

213. "Analogical Inference in Hume's Philosophy of Religion", *Faith and Philosophy*, 2, 1985, 287-294.

214. . "Bosanquet's Concept of Difficult Beauty", *The Journal of Aesthetics and Art Criticism*, 63, 1984, 79-87.

215. "Meinong's Theory of Defective Objects", *Grazer Philosophische Studien*, 15, 1982, 1- 19.

216. "Roland Barthes on the Aesthetics of Photography", *The Journal of the Theory and Criticism of the Visual Arts*, 1, 1982, 17-32.

4 Contributions to Books

1. "Cracking the Hard Problem of Consciousness", in *The Bloomsbury Companion to the Philosophy of Consciousness*, edited by Dale Jacquette (Bloomsbury, 2018), 258- 286.

2. "Wittgenstein and Schopenhauer", *A Companion to Wittgenstein*, edited by Hans-Johann Glock and John Hyman (Hoboken: John Wiley & Sons, Ltd., 2017), 59-73.

3. "Tractatus Objects and the Logic of Color Incompatibility", *Colours in the Development of Wittgenstein's Philosophy*, edited by Marcos Silva, (Palgrave Macmillan, 2017), 57-94.

4. "Tarski, Alfred", "Fractals", "Modality", *Handbook of Mereology*, edited by Hans Burkhardt and Guido Imaguire (Philosophia Verlag, 2017).

5. "Dummett on Truth Conditions, Frege's Analysis of Sentential Meaning, and the Slingshot Argument" *Truth, Meaning, Justification, and Reality: Themes from Dummett*, edited by Michael Frauchiger (Walter De Gruyter, 2017), 81-102.

6. "Chisholm and Brentano", *Handbook of Brentano and the Brentano School*, edited by Uriah Kriegel (Routledge, 2017), 358-364.

7. "Leibniz's Empirical, Not Empiricist Methodology", in *Leibniz-Scientist, Leibniz Philosopher*, edited by Lloyd Strickland and Julia Weckend (Palgrave-Macmillan, 2017), 179-202.

8. "Socratic Irony in Twain's Skeptical Religious Jeremiads", *Mark Twain and Philosophy*, edited by Alan Goldman, Great Authors and Philosophy Series (Rowman & Littlefield, 2017), 137-148.

9. "Anti-Meinongian Actualist Meaning of Fiction in Kripke's 1973 John Locke Lectures", *Existence, Fiction, Assumption: Meinongian Themes and the History of Austrian Philosophy*, edited by Mauro Antonelli and Marian David, Meinong Studies 6 (Berlin/Boston: Walter de Gruyter, 2016), 59-98.

10. "Representations of Evil", *The History of Evil in in Antiquity: 2000 BCE-450CE*, v.1, edited by Tom Angier, Chad Meister and Charles Taliaferro, (The History of Evil, 6 volumes), (Taylor & Francis Ltd., Routledge, 2016).

11. "9 Dale Jacquette", *Philosophy of Logic: 5 Questions*, edited by Thomas Adajian and Tracy Lupher (Automatic Press / VIP, 2016).

12. "Universal Logic or Logics in Resemblance Families", *The Road to Universal Logic Festschrift for 50th Birthday of Jean-Yves Beziau*, edited by Arnold Koslow and Arthur Buchsbaum, Studies in Universal Logic series (Basel: Birkhäuser (Springer Verlag), 2015), 309-325.

13. "Margolis on the Progress of Pragmatism", *Nordic Studies in Pragmatism* online book series, edited by Dirk-Martin Grube and Rob Sinclair (Helsinki: Nordic Pragmatism Network, 2015), 52-75.

14. "Domain Comprehension in Meinongian Object Theory", *Objects and Pseudo-Objects: Ontological Deserts and Jungles from Brentano to Carnap*, edited by

Bruno Leclercq, Sébastien Richard and Denis Seron (Boston-Berlin: Walter de Gruyter, 2014), 101-121.

15. "Austin on Conceptual Polarity and Sensation Deception Metaphors", *Austin on Language*, edited by Brian Garvey, Philosophers in Depth series (New York: Palgrave-Macmillan, 2014), 177-196.

16. "Practitions in Castañeda's Deontic Logic", *Essays on the Philosophy of Hector-Neri Castañeda*, edited by Adriano Palma (Boston/Berlin: Walter de Gruyter (Ontos), 2014), 29-46.

17. "Realism versus Idealism in the Nature-Nurture Dispute", *Defending Realism: Ontological and Epistemological Investigations*, edited by Guido Bonino, Greg Jesson, and Javier Cumpa, Foundations of Ontology series 7 (Boston/Berlin: Walter de Gruyter (Ontos), 2014), 401-415.

18. "Arthur Schopenhauer", *Oxford Bibliographies*, Oxford University Press (electronic format), http://www.oxfordbibliographies.com/view/document/, 2014 (17 printed pages).

19. "Semantic Intentionality and Intending to Act", in *An Anthology of Philosophical Studies*, 8, edited by Patricia Hanna (Athens: ATINER, 2014), 73-84. Previously, in electronic format, Athens Institute for Education and Research, Conference Paper Series, No: PHI2013-0616, 2013.

20. "Morally Tragic Life", *The International Encyclopedia of Ethics*, edited by Hugh LaFollette (Oxford: Wiley-Blackwell, 2013), 3439-3443. DOI: 10.1002/9781444367072.wbiee182.

21. "Belief State Intensity", *New Essays on Belief: Constitution, Content, and Structures,* edited by Nikolaj Nottelmann (Houndmills, Basingstoke: Palgrave-Macmillan, 2013), 209-229.

22. "Socioeconomic Darwinism from a South Park Perspective", *The Ultimate South Park and Philosophy: Respect My Philosophah!*, edited by Robert Arp and Kevin S. Decker (Malden: Wiley-Blackwell, 2013), 164-174.

23. "So, You Want to be a Mad ScientistÉ", *Frankenstein and Philosophy: The Shocking Truth*, edited by Nicholas Michaud (Chicago: Open Court Publishing, 2013), 37-45.

24. "Introducción: ?Qué es el cannabis y cómo lo podemos conseguir?", *Cannabis: ?De qué estamos hablano?* (2013 Spanish translation of Jacquette 2010 Wiley-Blackwell), 21-36.

25. "Navegación del espacio interior creativo con los innocentes placeres del hachis", *Cannabis: ?De qué estamos hablano?* (2013 Spanish translation of Jacquette 2010 Wiley-Black-well), 133-147.

26. "Schopenhauer's Philosophy of Logic and Mathematics", *Companion to Schopenhauer*, edited by Bart Vandenabeele (Oxford: Blackwell Publishing, 2012), 43-59.

27. "Die cartesianischen Argumente für die Unterscheidung von Körper und Geist", "Saul A. Kripkes Argument für einen Eigenschaftsdualismus von Körper und Geist" (German translations by Michael A. Conrad), *Die 100 wichtigsten philosophischen Argumente, herausgegeben von Michael Bruce und Steven Barbone* (Darmstadt: Wissenschaftliche Buchgesellschaft, 2012), 290-296; 301-303.

28. "Thinking Outside the Square of Opposition Box", *Around and Beyond the Square of Opposition*, edited by Jean-Yves Beziau and Dale Jacquette (New York: Birkhauser, Springer Verlag, 2012), 73-92.

29. "Applied Mathematics in the Sciences", *Between Logic and Reality: Modeling Inference, Action and Understanding*, edited by Majda Trobok, Nenad Miscevic, and Berislav Zarnic (Logic, Epistemology, and the Unity of Science, 25) (Dordrecht: Springer Verlag, 2012), Chapter 3, 29-58 (reprinted from The Croatian Journal of Philosophy, 2006).

30. "Collective Intentionality in the Theory of Meaning", *Intentionality: Historical and Systematic Perspectives*, edited by Alessandro Salice (Munich: Philosophia Verlag, 2012), 317-348.

31. "Cantor's Diagonalization and Turing's Cardinality Paradox", *Computing, Philosophy and the Question of Bio-Machine Hybrids (AISB/IACAP World Congress, Computing, Philosophy and the Question of Bio-Machine Hybrids*, edited by J.M. Bishop and Y.J. Erden (London: Society for the Study of Artificial Intelligence and Simulation of Behaviour (http://www.aisb.org.uk; ISBN 978-1-908187-11-6), 2012, 21-23.

32. "Brentano on Aristotle's Categories: First Philosophy and the Manifold Senses of Being", *Franz Brentano's Metaphysics and Psychology: Upon the Sesquicen-

tennial of Franz Brentano's Dissertation, edited by Ion Tanascescu (Bucharest: Zeta Books, 2012), Chapter 2, 53-94 (reprinted from Revue roumaine de Philosophie, 2011).

33. "The Sickness Unto Barfing", *The Catcher in the Rye and Philosophy: A Book for Bastards, Morons, and Madmen*, edited by Keith Dromm and Heather Salter (LaSalle: Open Court Publishing, 2012), 113-126.

34. "Hume's Enlightenment Aesthetics and Philosophy of Mathematics", *Hume and the Enlightenment*, edited by Craig Taylor and Stephen Buckle (London: Pickering & Chatto, 2011), 65-76.

35. "Hartmann's Philosophy of Mathematics", *The Philosophy of Nicolai Hartmann*, edited by Roberto Poli, Carlo Scognamiglio, and Frederic Tremblay (Berlin: Walter de Gruyter, 2011), 269-288.

36. "Evolutionary Emergence of Intentionality and Imagination", *A New Book of Nature: Philosophical Essays on the Imagination, Nature and God*, edited by Charles Taliaferro and Jil Evans (Oxford: Oxford University Press, 2011), 67-90.

37. "Descartes's Arguments for the Mind-Body Distinction", "Kripke's Argument for Mind- Body Property Dualism", *Just the Arguments: 100 of the Most Important Arguments in Western Philosophy*, edited by Michael Bruce and Steven Barone (Oxford: Wiley- Blackwell, 2011), 290-296; 301-303.

38. "Mind-Body Problem", "Reduction", "Supervenience", Glossary, *The Continuum Companion to Philosophy of Mind*, edited by James Garvey (London: Continuum Books, 2011), 299-301; 308-309; 312-313.

39. "Measure for Measure? Wittgenstein on Language-Game Criteria and the Paris Standard Metre Bar", *Wittgenstein's Philosophical Investigations: A Critical Guide*, edited by Arif Ahmed (Cambridge: Cambridge University Press, 2010), 49-65.

40. "Wittgenstein as Trans-Analytic-Continental Philosopher", *Postanalytic and Metacontinental: Crossing Philosophical Divides*, edited by James Williams, Jack Reynolds, James Chase, and Edwin Mares (London: Continuum Books, 2010), 157- 172.

41. "Developments in Philosophy of Science and Mathematics", *Nineteenth Century Philosophy: Revolutionary Responses to the Existing Order, The History*

of Continental Philosophy, edited by Alan D. Schrift and Daniel W. Conway, 8 vols., vol. 2 (Durham: Acumen Publishing Limited, 2010), 193-216; revised paperback edition, 2013.

42. "Introduction: Logical Possibilities and the Concept of a Logically Possible World" (with Guido Imaguire), in *Possible Worlds: Logic, Semantics and Ontology*, edited by Guido Imaguire and Dale Jacquette (Munich: Philosophia Verlag, 2010), 15-22.

43. "Kripkean Epistemically Possible Worlds", in *Possible Worlds: Logic, Semantics and Ontology*, edited by Guido Imaguire and Dale Jacquette (Munich: Philosophia Verlag, 2010), 99-140.

44. "Supervenience (on Steroids) and the Mind", *Causality and Motivation*, edited by Roberto Poli (Frankfurt: Ontos Verlag, 2010), 65-74.

45. "Journalism Ethics as Truth Telling in the Public Interest", *The Routledge Companion to News and Journalism*, edited by Stuart Allan (New York: Routledge, 2010; revised edition, 2012), 213-222.

46. "Schopenhauer, Arthur (1788-1860)", *A Dictionary of Philosophy of Religion*, edited by Charles Taliaferro and Elsa J. Marty (London: Continuum Books, 2010), 209-210.

47. "Proposition", "Soundness", "Type Theory", *Key Terms in Logic*, edited by Federica Russo and Jon Williamson (London: Continuum Books, 2010), 84-85; 93-94; 105- 106.

48. "Zombie Gladiators", *Zombies, Vampires, and Philosophy: New Life for the Undead*, edited by Richard Greene and K. Silem Mohammad (reprinting in expanded edition originally published as The Undead and Philosophy: Chicken Soup for the Soulless, 2006 below) (Chicago and LaSalle: Open Court Press, 2010), 105-118.

49. "Zombies als Gladiatoren", *Die Untoten und die Philosophie: Schlauer werden mit Zombies, Werwölfen und Vampiren*, edited by Richard Greene and K. Silem Mohammad (German translation of "Zombie Gladiators" in Zombies, Vampires, and Philosophy: New Life for the Undead, 2010 above, and in The Undead and Philosophy: Chicken Soup for the Soulless, 2006 below) (Stuttgart: Klett-Cotta, 2010), 141-160.

50. "Introduction: What is Cannabis and How Can We Get Some?", *Cannabis & Philosophy: What Were We Just Talking About?* Edited by Dale Jacquette (Oxford, Malden: Wiley-Blackwell, 2010), 1-17 (Preface, xv-xix).

51. "Navigating Creative Inner Space on the Innocent Pleasures of Hashish", *Cannabis & Philosophy: What Were We Just Talking About?* Edited by Dale Jacquette (Oxford, Malden: Wiley-Blackwell, 2010), 121-136.

52. "Meditations on Meinong's Golden Mountain", *Russell versus Meinong: The Legacy of 'On Denoting'*, edited by Nicholas Griffin and Dale Jacquette (New York: Routledge, 2009), 169-203.

53. "Deductivism and the Informal Fallacies", *Pondering on Problems of Argumentation: Twenty Essays on Theoretical Issues*, edited by Frans H. van Eemeren and Bart Garssen (Berlin: Springer Verlag, 2009), 105-114.

54. "Nicholas Rescher's Systematic Philosophy", *Introduction to Reason, Method, and Value: A Reader on the Philosophy of Nicholas Rescher*, edited and with a critical introduction by Dale Jacquette (Heusenstamm: Ontos Verlag, 2009), 1-17.

55. "Logic for Meinongian Object Theory Semantics", *Handbook of the History and Philosophy of Logic*, edited by Dov M. Gabbay and John Woods, Volume V: Logic from Russell to Gödel (Amsterdam: Elsevier Science (North-Holland), 2009), 29-76.

56. "Justification in Ethics", *Morality and Politics: Reading Boylan's A Just Society*, edited by John-Stewart Gordon (Rowman & Littlefield, Lexington Books, 2009), 55-69.

57. "Boole's Logic", *Handbook of the History of Logic*, Volume 4: British Logic in the Nineteenth Century, edited by Dov M. Gabbay and John Woods (Amsterdam: North- Holland (Elsevier Science), 2008), 331-379.

58. "Designation", "Knowledge: Probable", "Scheffler, Israel", "Synthesis", American *Philosophy: An Encyclopedia*, edited by John Lachs and Robert Talisse (New York: Routledge, 2008); 172-173; 442-444; 698; 746-747.

59. "Reasoning Awry: An Introduction to Woods and Walton, Fallacies: Selected Papers 1972-1982", Foreword to John Woods and Douglas Walton, *Fallacies: Selected Papers 1972-1982* (London: King's College Publications, 2007), vii-xvi.

60. "Introduction: Philosophy of Logic Today", *Handbook of the Philosophy of Logic*, edited by Dale Jacquette (Amsterdam: North-Holland (Elsevier Science), 2007), 1-12.

61. "On the Relation of Informal to Symbolic Logic", *Handbook of the Philosophy of Logic*, edited by Dale Jacquette (Amsterdam: North-Holland (Elsevier Science), 2007), 131-154.

62. "Thirst for Authenticity: An Aesthetics of the Brewer's Art", *Beer and Philosophy: The Unexamined Beer Isn't Worth Drinking*, edited by Steven D. Hales (Oxford: Blackwell Publishing, 2007), 15-30.

63. "Crossroads of Logic and Ontology: A Modal-Combinatorial Analysis of Why There is Something Rather Than Nothing", *Essays in Logic and Ontology, Dedicated to Jerzy Perzanowski*, edited by Jacek Malinowski and Andrzej Pietruszczak (Amsterdam: Rodopi, Pozman Studies in Philosophy and the Humanities 91, 2007), 17-46.

64. "Preface: After Russell, After Meinong", *After 'On Denoting': Themes from Russell and Meinong* (edited with Nicholas Griffin and Kenneth Blackwell) (Hamilton: McMaster University for The Bertrand Russell Research Centre, 2007), 5-6 (also published as special issue of Russell: The Journal of Bertrand Russell Studies, 2007, above).

65. "Deductivism and the Informal Fallacies", *Proceedings of the Sixth Conference of the Inter-national Society for the Study of Argumentation (ISSA)*, edited by Frans van Eemeren, J. Anthony Blair, Charles A. Willard, and Bart Garssen (Amsterdam: Sic Sat (International Center for the Study of Argumentation), 2007, vol. 1, 687-692.

66. "Hume on the Infinite Divisibility of Extension and Exact Geometrical Values", *New Essays on David Hume*, edited by Emilio Mazza and Emanuele Rochetti, Filosofia e Scienza nell'età Moderna (Milan: FrancoAngeli, 2007), 81-100.

67. "Satan Lord of Darkness in South Park Cosmology",*South Park and Philosophy: You Know, I Learned Something Today*, edited by Robert Arp (Oxford: Blackwell Publishing, 2007), 250-262.

68. "Animadversions on the Logic of Fiction and Reform of Modal Logic", *Mistakes of Reason: Proceedings of a Conference in Honour of John Woods*, edited by Kent A. Peacock and Andrew D. Irvine (Toronto: University of Toronto Press, 2006), 49-63.

69. "Tarski's Analysis of Logical Consequence and Etchemendy's Criticism of Tarski's Modal Fallacy", *The Lvov-Warsaw School: The New Generation, Poznan Studies in the Philosophy of the Sciences and Humanities 89*, edited by Jacek Juliusz Jadacki and Jacek Pasniczek (Amsterdam: Rodopi Editions, 2006), 345-368.

70. "Dale Jacquette", *Masses of Formal Philosophy: Aim, Scope, Direction*, edited by Vincent F. Hendricks and John Symons (New York: Automatic Press / VIP, 2006), 59-81.

71. "Twardowski, Brentano's Dilemma, and the Content-Object Distinction", *Actions, Products, and Things: Brentano and Polish Philosophy*, Mind and Phenomenology series, edited by Arkadiusz Chrudzimski and Dariusz Lukasiewicz (Frankfurt: Ontos Verlag, 2006), 9-33.

72. "Zombie Gladiators", *The Undead and Philosophy: Chicken Soup for the Soulless*, edited by Richard Greene and K. Silem Mohammad (Chicago and LaSalle: Open Court Press, 2006), 105-118.

73. "Wittgensteins 'Tractatus' und die Logik der Fiktion", *Wittgenstein und die Literatur*, edited by John Gibson and Wolfgang Huemer, translated by Martin Suhr (Frankfurt am Main: Surkamp Verlag, 2006), 448-467 (original English version published by Routledge 2004 below).

74. "Brentano, Franz", "Husserl, Edmund", *Europe: Encyclopedia of Industry and Empire 1789-1914, Scribner's Library of Modern Europe*, edited by John Merriman and Jay Winter (New York: Gale/Thomson, Charles Scribner's Sons, 2006), vol. 1 (Abdul Hamid II to Colonialism), 298-300; vol. 2 (Colonies to Huysmans), 1099-1101.

75. "Idealism: Schopenhauer, Schiller and Schelling", *The Routledge Companion to Aesthetics*, second edition, edited by Berys Gaut and Dominic McIver Lopes (London: Routledge, 2005), 83-95.

76. "Kripke's Modal Objection to the Description Theory of Reference", *We Will Show Them! Essays in Honour of Dov Gabbay*, edited by Sergei Artemov, Howard Barringer, Artur d'Avila Garcez, Luis C. Lamb and John Woods (London: Kings College Publications, 2005), 143-168.

77. "Arthur Schopenhauer: The World as Will and Representation", *Central Works of Philosophy, 3*, The Nineteenth Century, edited by John Shand (Chesham: Acumen Books, 2005), 93-126.

78. "Two Sides of Any Issue", *The Uses of Argument: Proceedings of a Conference at McMaster University 18-21 May 2005*, edited by David Hitchcock with Daniel Farr (St. Catherines: Ontario Society for the Study of Argumentation, 2005), 209-217.

79. "Introduction: Brentano's Philosophy", *The Cambridge Companion to Brentano*, edited by Dale Jacquette (Cambridge: Cambridge University Press, 2004), 1-19.

80. "Brentano's Concept of Intentionality", *The Cambridge Companion to Brentano*, edited by Dale Jacquette (Cambridge: Cambridge University Press, 2004), 98-130.

81. "Wittgenstein's Tractatus and the Logic of Fiction", *The Literary Wittgenstein*, edited by John Gibson and Wolfgang Huemer (London: Routledge, 2004), 305-317. Translated into Chinese (Mandarin) (Shanghai Sanhui Press Ltd, 2009), 404-422.

82. "Wittgenstein on Lying as a Language Game", *The Third Wittgenstein: The Post- Investigations Works*, edited by Daniéle Moyal-Sharrock (Aldershot: Ashgate

83. "Assumption and Mechanical Simulation of Hypothetical Reasoning", *Phenomenology and Analysis: Essays on Central European Philosophy*, edited by Arkadiusz Chrudzimski and Wolfgang Huemer (Frankfurt/Lancaster: Ontos Verlag, 2004), 323-358.

84. "Monod, Jacques (Lucien)", *Encyclopedia of Modern French Thought*, edited by Christopher John Murray (New York: Taylor & Francis, 2004), 488-490.

85. "Kripke on Identity and the Description Theory of Reference", *The Logica Yearbook 2002*, edited by Timothy Childers and Ondrej Majer (Prague: Filosofia, 2003) (Institute of Philosophy, Academy of Sciences of the Czech Republic), 109-116.

86. "Introduction: Psychologism the Philosophical Shibboleth", *Philosophy, Psychology, and Psychologism: Critical and Historical Readings on the Psychological Turn in Philosophy* (edited) (Kluwer Academic Publishing / Philosophical Studies Series, volume 91, 2003), 1-19; reprinted from Philosophy and Rhetoric (see above).

87. "Psychologism Revisited in Logic, Metaphysics, and Epistemology", *Philosophy, Psychology, and Psychologism: Critical and Historical Readings on the Psychological Turn in Philosophy* (edited) (Kluwer Academic Publishing / Philosophical Studies Series, volume 91, 2003), 245-262; reprinted from Metaphilosophy (see above).

88. "Rescher on the Functionalist Pragmatic Roots of Reason", *Foreword to Nicholas Rescher, Rationality in Pragmatic Perspective* (Edwin Mellen Press, 2003), i-iv.

89. "David Lewis on Meinongian Logic of Fiction", *Writing the Austrian Traditions: Relations Between Philosophy and Literature*, edited by Wolfgang Huemer and Marc-Oliver Schuster (Toronto: University of Toronto Press, Wirth-Institute for Austrian and Central European Studies, University of Alberta, 2003), 101-119. Electronic format publication: http://www.unipr.it/ hue-wol48/wat.html

90. "Socrates on Rhetoric, Truth, and Courtroom Argumentation in Plato's Apology", *Informal Logic at 25: Proceedings of the Windsor Conference*, edited by J. Anthony Blair, Daniel Farr, Hans V. Hansen, Ralph H. Johnson, and Christopher W. Tindale (Windsor: Ontario Society for the Study of Argumentation, 2003), CD-ROM.

91. "Diagonalization in Logic and Mathematics", *Handbook of Philosophical Logic*, 2nd Edition, Volume 11, edited by Dov M. Gabbay and Franz Guenthner (Dordrecht: Kluwer Academic Publishing, 2002), 55-147.

92. "Wittgenstein on Thoughts as Pictures of Facts and the Transcendence of the Metaphysical Subject", *Wittgenstein and the Future of Philosophy: A Reassessment After 50 Years / Wittgenstein und die Zukunft der Philosophie. Eine Neubewertung nach 50 Jahren*, edited by Rudolf Haller and Klaus Puhl (Vienna: öbv& Hpt, 2002), 160-170.

93. "Wittgenstein, Ludwig", *Encyclopedia of Communication and Information*, edited by Jorge Reina Schement (New York: Macmillan Reference, 2002), 1090-1092.

94. "Logic, Philosophy, and Philosophical Logic", *A Companion to Philosophical Logic*, edited by Dale Jacquette (Blackwell Publishers, 2002), 1-8.

95. "Modality of Deductively Valid Inference", *A Companion to Philosophical Logic*, edited by Dale Jacquette (Blackwell Publishers, 2002), 256-261.

96. "Logic and Philosophy of Logic", *Philosophy of Logic: An Anthology*, edited by Dale Jacquette (Blackwell Publishers, 2002), 1-6. "Introduction to Part I" (Classical Logic), 9-11; "Introduction to Part II" (Truth, Propositions, and Meaning), 55-57; "Introduction to Part III" (Quantifiers and Quantificational Theory), 143-145; "Introduction to Part IV" (Validity, Inference, and Entailment), 201-204; "Introduction to Part V" (Modality, Intensioinality, and Propositional Attitude), 271- 274.

97. "Mathematics and Philosophy of Mathematics", *Philosophy of Mathematics: An Anthology*, edited by Dale Jacquette (Blackwell Publishers, 2002), 1-10. "Introduction to Part I" (The Realm of Mathematics), 13-17; "Introduction to Part II" (Ontology of Mathematics and the Nature and Knowledge of Mathematical Truth), 85-89; "Introduction to Part III" (Models and Methods of Mathematical Proof), 165-

98. ; "Introduction to Part IV" (Intuitionism), 261-267; "Introduction to Part V" (Philosophical Foundations of Set Theory), 337-343.

99. "A History of Early Analytic Philosophy of Language", *Analytic Philosophy: Classic Readings*, edited by Steven B. Hales (Wadsworth Publishing, 2001), 11-20.

100. "Introduction: Meinong in His and in Our Times" (with Liliana Albertazzi and Roberto Poli), *The School of Alexius Meinong*, edited by Liliana Albertazzi, Dale Jacquette, and Roberto Poli (Ashgate Publishing, 2001), 3-48.

101. . "Außersein of the Pure Object", *The School of Alexius Meinong*, edited by Liliana Albertazzi, Dale Jacquette, and Roberto Poli (Ashgate Publishing, 2001), 373-396.

102. "Nuclear and Extranuclear Properties", *The School of Alexius Meinong*, edited by Liliana Albertazzi, Dale Jacquette, and Roberto Poli (Ashgate Publishing, 2001), 397-426.

103. "Of Time and the River in Kant's Refutation of the Problematic Idealism", *Kant und die Berliner Aufklarung: Akten des Neunten Internationalen Kant-Kongresses*, edited by Volker Gerhardt, Rolf-Peter Horstmann and Ralph Schumacher (Berlin: Walter de Gruyter, 2001), 5 volumes, volume 3, 571-582.

104. "The Deconstruction Debacle in Theory and Practice", *Proceedings of the Twentieth World Congress of Philosophy*, vol. 8, 'Contemporary Philosophy',

edited by Daniel O. Dahlstrom (Bowling Green: Philosophy Documentation Center, 2000), 67-79.

105. "On Certainty (Wittgenstein, Ludwig)", *World Philosophers and Their Works*, revised edition, 3 vols., edited by John K. Roth (Pasadena: Salem Press, 2000), vol. 3, 2002- 2005.

106. "Schopenhauer on Death", *The Cambridge Companion to Schopenhauer*, edited by Christo-pher Janaway (Cambridge: Cambridge University Press, 1999), 293-317.

107. "Margolis and the Metaphysics of Culture", *Interpretation, Relativism, and the Metaphysics of Culture: Themes in the Philosophy of Joseph Margolis*, edited by Michael Krausz and Richard Shusterman (Amherst: Humanity Books (Prometheus), 1999), 225-261.

108. "Goethe, Johann Wolfgang von", *Encyclopedia of Aesthetics*, 4 vols., edited by Michael Kelly (Oxford: Oxford University Press, 1998), vol. 2, 311-315; revised for 2nd edition.

109. "On the Relation of Informal to Formal Logic", *Argumentation and Rhetoric*, edited by Hans V. Hansen, Christopher W. Tindale, and Athena V. Colman (St. Catharines: Ontario Society for the Study of Argumentation, 1998), CD-ROM.

110. "Ignorance is No Excuse (for Deductively Invalid Inference)", *Argumentation and Rhetoric*, edited by Hans V. Hansen, Christopher W. Tindale, and Athena V. Colman (St. Catharines: Ontario Society for the Study of Argumentation, 1998), CD-ROM.

111. "Haller on Wittgenstein and Kant", *Austrian Philosophy Past and Present: Essays in Honor of Rudolf Haller*, edited by Keith Lehrer and Johannes Christian Marek, Boston Studies in the Philosophy of Science (Dordrecht: Kluwer Academic Publishers, 1997), 29-44.

112. "A Priori / A Posteriori Distinction", "Analytic/Synthetic Distinction", "Kant, Immanuel", "Sensationalism", *Encyclopedia of Empiricism*, edited by Don Garrett and Edward Barbanell (Westport: Greenwood Press, 1997), 1-2; 7-10; 197-202; 388-390.

113. "Constructibility and the Analysis of Quantifiers in Wittgenstein's Tractatus", *The Role of Pragmatics in Contemporary Philosophy: Contributions of*

the Austrian Ludwig Wittgenstein Society, 20th International Wittgenstein Symposium, edited by Paul Weingartner, Gerhard Schurz and Georg Dorn (Kirchberg am Wechsel: Austrian Ludwig Wittgenstein Society, 1997), 5, vol. I, 419-426.

114. "Alexius Meinong (1853-1920)", *The School of Franz Brentano*, edited by Liliana Albertazzi, Massimo Libardi, and Roberto Poli (Dordrecht: Kluwer Academic Publish-ers, 1996), 131-159.

115. "Schopenhauer's Metaphysics of Appearance and Will in the Philosophy of Art", in *Schopenhauer, Philosophy, and the Arts*, edited by Dale Jacquette (Cambridge: Cambridge University Press, 1996), 1-36.

116. "Hume, David", "Quine, Willard Van Orman", *Encyclopedia of Rhetoric and Composition: Communication from Ancient Times to the Information Age*, edited by Theresa Enos (New York and London: Garland Publishing, 1996), 332-333; 580- 581.

117. "Introduction: Findlay and Meinong", *J.N. Findlay, Meinong's Theory of Objects and Values*, Edited with an Introduction by Dale Jacquette (Gregg Revivals) (Aldershot: Ashgate Publishing, 1995), xxv-lvi.

118. "Abstract Entity", "Brentano, Franz", "Extensionalism", "Haecceity", "Impredicative Defini-tion", "Meinong, Alexius", "Subject-Object Dichotomy", "Use-Mention Distinction", *The Cambridge Dictionary of Philosophy*, edited by Robert Audi (Cambridge: Cambridge University Press, 1995), 3-4; 86-87; 258; 308; 363-364; 477-478; 773-774; 823-824; (second edition, 1999), 3-4; 100-101; 300; 359-360; 421; 551-553; 885-886; 942; "Act-Object Psychology", 9.

119. "Wittgenstein's Private Language Argument and Reductivism in the Cognitive Sciences", *Philosophy and the Cognitive Sciences*, edited by Roberto Casati, Barry Smith, and Graham White (Vienna: Hölder-Pichler-Tempsky, 1994), 89-99.

120. "Desire", "Intention", "Responsibility", "Wittgenstein, Ludwig", Ethics (Magill Ready Reference), edited by John K. Roth (Pasadena: Salem Press, 1994), 3 vols., 221- 222; 444-445; 742-746; 936-937.

121. "Buridan's Bridge", *Zeit und Zeichen*, edited by Tilman Borsche, Johann Kreuzer, Helmut Pape, and Günter Wohlfart, Schriften der Académie du Midi (Munich: Wilhelm Fink Verlag, 1993), 227-241; reprinted from Philosophy 1991 (above).

122. "Meinongian Models of Scientific Law", *Theories of Objects: Meinong and Twardowski*, edited by Jacek Pasniczek (Lublin: Wydawnictwo Uniwersytetu Marii Curie- Sklodows-kiej, 1992), 86-104.

123. "Self-Reference and Infinite Regress in Philosophical Argumentation" (with Henry W. Johnstone, Jr.), *Argumentation Illuminated*, edited by Frans H. van Eemeren, Rob Grootendorst, J. Anthony Blair, and Charles A. Willard (Amsterdam: International Society for the Study of Argumentation, 1992), 122-130.

124. "The Myth of Pure Syntax", *Topics in Philosophy and Artificial Intelligence*, edited by Liliana Albertazzi and Roberto Poli (Bolzano: Instituto Mitteleuropeo di Cultura, 1991), 1-14.

125. "Categorical Moral Maxims in Kant's Categorical Imperative", *Akten des Siebenten Internationalen Kant-Kongresses*, edited by Gerhard Funke (Bonn: Bouvier Verlag, 1991), vol. II.1, 313-322.

126. "Validity, Self-Reference, and the Pseudo-Scotus Pseudo-Paradox", *Proceedings of the Second International Conference on Argumentation*, edited by Frans H. van Eemeren, Rob Grootendorst, J. Anthony Blair, and Charles A. Willard (Amsterdam: International Society for the Study of Argumentation, 1991), vol. 1A, 211-217.

127. "Definite Descriptions", "Extensionalism", "Lambda Abstraction", "Mally, Ernst", *Handbook of Metaphysics and Ontology*, edited by Hans Burkhardt and Barry Smith (Munich and Vienna: Philosophia Verlag, 1991), 2 vols., 201-202; 268-269; 435; 485-486.

128. "Metaphilosophy in Wittgenstein's City", *Ludwig Wittgenstein: A Symposium on the Centennial of his Birth*, edited by Souren Tegharian, Anthony Serafini and Edward M. Cook (Wakefield: Longwood Academics, 1990), 31-42; reprinted in International Studies in Philosophy 1993 (above).

129. "Moral Value and the Sociobiological Reduction", *Inquiries Into Values: The Inaugural Session of the International Society for Value Inquiry, Problems in Contemporary Philosophy, XI*, edited by Sander H. Lee (Lewiston: The Edwin Mellon Press, 1988), 685-694.

130. "Meinongian Mathematics and Metamathematics", *Logic, Philosophy of Science and Epistemology*, Proceedings of the Eleventh International Wittgenstein Symposium (Vienna: Hölder-Pichler-Tempsky, 1987), 109-112.

131. "Turnstile Operations in Three-Valued Logic", *Proceedings of the Fifteenth International Symposium on Multiple-Valued Logic*, Institute of Electrical and Electronics Engineers (IEEE), Technical Committee on Multiple-Valued Logic (Silver Spring: Computer Soc-iety Press, 1985), 45-48.

132. "Voltaire and the Philosophes", *The History of Evil*, 6 volumes, edited by Chad Meister and Charles Taliaferro, IV, The History of Evil in the 18th and 19th Centuries: 1700- 1900, v. 4, edited by Douglas Hedley, (Taylor & Francis, Routledge, announced publication, June 2018).

133. "Logic of Identity", *Internet Encyclopedia of Philosophy (IEP)*, electronic format, forthcom-ing.

134. "Brentano on Aristotle's Psychology of the Active Intellect", *Aristotelische Forschungen im 19. Jahrhundert*, edited by Gerald Hartung, Colin Guthrie King, and Christof Rapp (Walter De Gruyter, announced publication, October 2020).

135. "Referential Analysis of Quotation", *Semantic and Pragmatic Aspects of Quotation*, edited by Paul Saka (Springer Verlag, announced publication, January 2018).

136. "Objectification in Schopenhauer's Metaphysics", *Oxford Handbook of Schopenhauer*, edited by Robert Wicks (Oxford University Press, anticipated 2019).

137. "Franz Brentano's Theodicy", *Wiley-Blackwell Encyclopedia of the Philosophy of Religion*, edited by Stewart Goetz and Charles Taliaferro (Wiley-Blackwell, anticipated publication 2020).

138. "Frege on Indirect Discourse", *Indirect Reports*, edited by Alessandro Capone, Vahid Parvaresh, Alessandra Falzone (Springer Verlag, forthcoming).

5 Writings Reprinted in Anthologies

1. "Roland Barthes on the Aesthetics of Photography", from The Journal of the Theory and Criticism of the Visual Arts, 1, 1982, in *Roland Barthes*, edited by Mike Gane and Nicholas Gane (London: Sage Publications Ltd., 2004), Sage Masters of Modern Social Thought Series, 3 vols., Part Ten: Themes (III) Photography / Camera Lucida / La Chambre Claire, 225-239.

2. "Wittgenstein and the Color Incompatibility Problem", from *Ludwig Wittgenstein: Critical Asssessments of Leading Philosophers*, Second Series, edited by by Stuart Shanker and David Kilfoyle (London: Routledge, 2002), vol. I, The Early Wittgenstein: From the Notebooks to the Philosophical Grammar, 204-218.

3. "Dualisms of Mental and Physical Phenomena", from *Philosophy of Mind, in Problems in Mind: Readings in Contemporary Philosophy of Mind*, edited by Jack S. Crumley II (Mountain View: Mayfield Publishing Company, 2000), 37-43.

4. "Schopenhauer on the Ethics of Suicide", from *Continental Philosophy Review*, in Nineteenth Century Literary Criticism, 157, edited by Russell Whittaker (Gale Group / Thomson Publishing, 2000), 33 1, 43-58.

5. "Schopenhauer's Circle and the Principle of Sufficient Reason", from *Metaphilosophy, in Nineteenth Century Literary Criticism*, 157, edited by Russell Whittaker (Gale Group / Thomson Publishing).

6 Reviews

1. Review of Mark van Atten, Essays on Gödel's Reception of Leibniz, Husserl, and Brouwer, *Phenomenological Reviews*, OPHEN 16.04.2016, (electronic format) http://reviews.ophen.org/2016/04/16/mark-van-atten-essays-godels-receptionleibniz- husserl-brouwer/.

2. Review of Ted Honderich, Actual Consciousness, *Notre Dame Philosophical Reviews* (elec-tronic format), http://ndpr.nd.edu/news/60148-actual consciousness/ Vd5hFSJGPSU; 2015.08.35.

3. Review of Jan Willem Wieland, Infinite Regress Arguments, *Argumentation*, 28, 2015, 351-360. Electronic Online First format, DOI: 10.1007/s10503-014-9338-y.

4. Review of Maurice A. Finocchiaro, Meta-Argumentation: An Approach to Logic and Argu-mentation Theory, *Argumentation*, 28, 2014, 221-230. Electronic Online First format, DOI: 10.1007/s10503-013-9301-3.

5. Review of Christopher Hookway, The Pragmatic Maxim: Essays on Peirce and Pragmatism, *Journal for the History of Analytical Philosophy*, 2, 2013, [1]-[6]. Electronic format, DOI: http://dx.doi.org/10.15173/jhap.v2i3.22).

6. Review of George Englebretsen, Robust Reality: An Essay in Formal Ontology, *Ratio*, 26, 2013, 106-114.

7. 'Schopenhauer as the World's Clear Philosophical Eye' (Review Essay of David E. Cartwright, Schopenhauer: A Life), *British Journal for the History of Philosophy*, 19, 2011, 983-996.

8. Review of John Russell Roberts, A Metaphysics for the Mob: The Philosophy of George Berkeley, *Faith and Philosophy*, 28, 2011, 468-472.

9. Review of Christopher Ryan, Schopenhauer's Philosophy of Religion, *Religious Studies*, 46, 2010, 545-551.

10. Review of Robin Rollinger, Austrian Phenomenology: Brentano, Husserl, Meinong and Others on Mind and Object, *Grazer Philosophische Studien*, 80, 2010, 317-322.

11. Review of Guillermo E. Rosado Haddock, The Young Carnap's Unknown Master: Husserl's Influence on Der Raum and Der Logische Aufbau Der Welt, History and *Philosophy of Logic*, 30, 2009, 194-200.

12. Review of Garry L. Hagberg, Describing Ourselves: Wittgensgtein and Autobiographical Consciousness, *Notre Dame Philosophical Reviews* (electronic format), http://ndpr.nd.edu/review.cfm?id=17525; 2009.09.27.

13. Review of Terrance W. Klein, Wittgenstein and the Metaphysics of Grace, *The Review of Metaphysics*, 62, 2009, 668-670.

14. Review of Dean W. Zimmerman, editor, *Oxford Studies in Metaphysics*, Volume 2, Grazer Philosophische Studien, 76, 2008, 279-285.

15. Review of Liliana Albertazzi, Immanent Realism: An Introduction to Brentano, *The Review of Metaphysics*, 62, 2008, 123-125.

16. Review of Colin McGinn, Shakespeare's Philosophy, *The Journal of Aesthetics and Art Criticism*, 65, 2007, 421-424.

17. Review of Donald M. Baxter, Hume's Difficulty: Time and Identity in the Treatise, *Hume Studies*, 33, 2007, 352-357.

18. Review of Alice Crary, editor, Wittgenstein and the Moral Life: Essays in Honor of Cora Diamond, *Notre Dame Philosophical Reviews* (electronic format), http://ndpr.nd.edu/ review.cfm?id=11863; 2007.12.05.

19. Review of Rom Harré and Michael Tissaw, Wittgenstein and Psychology: A Practical Guide, *Journal of the History of Philosophy*, 45, 2007, 169-170.

20. Critical Notice of Danielle Macbeth, Frege's Logic, *Canadian Journal of Philosophy*, 36, 2006, 609-632.

21. Review of Paolo Mancosu, Klaus Frovin J£rgensen, and Stig Andur Pedersen, editors, Visualization, Explanation and Reasoning Styles in Mathematics, *The Mathematical Intelligencer*, 28, 2006, 79-81. Electronic version available from SpringerLink at http://www.springerlink.com/content/w213543j06041577/.

22. Review of Hans-Johann Glock, Quine and Davidson on Language, Thought and Reality, *Philosophical Investigations*, 29, 2006, 97-103.

23. Review of Anna Sierzulska, Meinong on Meaning and Truth, *Notre Dame Philosophical Reviews* (electronic format), http://ndpr.nd.edu/review.cfm?id=5781; 2006.02.17.

24. Review of Edmund Husserl, Philosophy of Arithmetic: Psychological and Logical Investigations with Supplementary Texts from 1887-1901, translated by Dallas Willard, *The Review of Metaphysics*, 59, 2005, 428-431.

25. Review of Gordon Baker, Wittgenstein's Method: Neglected Aspects, *International Philosophical Quarterly*, 45, 2005, 264-266.

26. Review of Karl Schuhmann, Selected Papers on Phenomenology, *Studia Phenomenologica*, 5, 2005, 388-390.

27. Review of Michael Rea, World Without Design: The Ontological Consequences of Naturalism, *Faith and Philosophy*, 21, 2004, 125-130.

28. "Mathematical Fiction and Structuralism in Chihara's Constructibility Theory" (Review of Charles S. Chihara, A Structural Account of Mathematics), *History and Philosophy of Logic*, 25, 2004, 319-324.

29. Review of Dov M. Gabbay and John Woods, Agenda Relevance: A Study in Formal Pragmatics (A Practical Logic of Cognitive Systems, Volume I), *Studia Logica: An International Journal for Symbolic Logic*, 77, 2004, 133-139.

30. Review of Sun-Joo Shin, The Iconic Logic of Peirce's Graphs, *Transactions of the Charles S. Peirce Society*, 39, 2003, 127-133.

31. Review of Oswald Hanfling, Wittgenstein and the Human Form of Life, *International Philosophical Quarterly*, 43, 2003, 384-387.

32. Review of Arkadiusz Chrudzimski, Intentionalitätstheorie beim frühen Brentano, *The Review of Metaphysics*, 46, 2002, 163-167.

33. Review of Jerry A. Fodor, The Mind Doesn't Work That Way: The Scope and Limits of Computational Psychology, *The Philosophers' Magazine*, 18, 2002, 59.

34. Review of Stephen Buckle, Hume's Enlightenment Tract: The Unity and Purpose of An Enquiry Concerning Human Understanding, *Hume Studies*, 28, 2002, 149-153.

35. Review of Katherine Hawley, How Things Persist, *International Philosophical Quarterly*, 42, 2002, 551-554.

36. Review of Laurence Goldstein, Clear and Queer Thinking: Wittgenstein's Development and his Relevance to Modern Thought, *Mind*, 110, 2001, 207-211.

37. Review of Christopher Williams, A Cultivated Reason: An Essay on Hume and Humeanism, *The British Journal for the History of Philosophy*, 9, 2001, 591-594.

38. "Light, Dark, and Shades of Gray Matter" [Review of Charles Don Keyes, Brain Mystery Light and Dark: The Rhythm and Harmony of Consciousness], *Metascience*, 10, 2001, 81-86.

39. Review of David Owen, Hume's Reason, *The British Journal for the History of Philosophy*, 9, 2001, 377-381.

40. Review of Christopher Janaway, Self and World in Schopenhauer's Philosophy, *The Review of Metaphysics*, 54, 2001, 660-661.

41. "Wagner, Philosophy, and the Apotheosis of Nineteenth-Century Opera" [Review of Bryan Magee, Wagner and Philosophy], *Wagner*, 22, 2001, 59-62.

42. Review of Christopher Janaway, editor, Willing and Nothingness: Schopenhauer as Nietzsche's Educator, *Philosophical Books*, 41, 2000, 184-186.

43. Review of Harold W. Noonan, Hume on Knowledge, *The Philosophers' Magazine*, 10, 2000, 55.

44. Review of J.N. Mohanty, Logic, Truth, and the Modalities: From a Phenomenological Perspective, *History and Philosophy of Logic*, 21, 2000, 172-173.

45. Review of Jacek Pasniczek, The Logic of Intentional Objects: A Meinongian Version of Classical Logic, *Journal of Symbolic Logic*, 64, 1999, 1847-1849.

46. Review of Carol Rovane, The Bounds of Agency: An Essay in Revisionary Metaphysics, *Philosophy in Review / Comptes rendus philosophiques*, 19, 1999, 55-57.

47. Review of Nicholas Rescher, Objectivity: The Obligations of Impersonal Reason, *Philosophy and Rhetoric*, 32, 1999, 286-291.

48. Review of Graham Priest, Beyond the Limits of Thought, *Informal Logic*, 19, 1999, 221- 226.

49. Review of Marie-Luise Schubert Kalsi, Alexius Meinong's Elements of Ethics, With a Trans-lation of the Fragment Ethische Bausteine, *The Review of Metaphysics*, 52, 1999, 727-730.

50. Review of H.E. Mason, editor, Moral Dilemmas and Moral Theory, *Philosophical Books*, 39, 1998, 62-65.

51. Review of László Pólos and Michael Masuch, editors, Applied Logic: How, What and Why (Logical Approaches to Natural Language), *Studia Logica: An International Journal for Symbolic Logic*, 60, 1998, 336-340.

52. Review of P.F. Strawson, Entity and Identity and Other Essays, *International Philosophical Quarterly*, 38, 1998, 322-323.

53. Review of Quentin Skinner, Reason and Rhetoric in the Philosophy of Hobbes, *Philosophy and Rhetoric*, 31, 1998, 74-79.

54. Essay Review of P.M.S. Hacker, Wittgenstein's Place in Twentieth-Century Analytic Philosophy, *History and Philosophy of Logic*, 18, 1997, 109-114.

55. "The Microscope in Early Modern Science and Philosophy" (Essay Review of Catherine Wilson, The Invisible World: Early Modern Philosophy and the Invention of the Micro-scope), *Studies in History and Philosophy of Science*, 28, 1997, 377-386.

56. Review of David Berman, George Berkeley: Idealism and the Man, *Religious Studies*, 31, 1995, 404-407.

57. Review of Hollibert E. Phillips, Vicissitudes of the I: An Introduction to the Philosophy of Mind, *The Personalist Forum*, 11, 1995, 55-58.

58. Review of Jaakko Hintikka, "Carnap's Work in the Foundations of Logic and Mathematics in a Historical Perspective", *Mathematical Reviews*, 95f, 1995, 3133.

59. Review of Andrés Rivadulla, "Wahrscheinlichkeitsaussagen, statistische Inferenz und Hypothesenwahrscheinlichkeit in L. Wittgensteins Schriften der Übergangsperiode", *Mathematical Reviews*, 95i, 1995, 5080.

60. Review of Nicholas Rescher, Pluralism: Against the Demand for Consensus, *Philosophical Books*, 35, 1994, 264-266.

61. Review of Robert Audi, Practical Reasoning, *Philosophy and Rhetoric*, 26, 1993, 85-89.

62. Review of Douglas N. Walton, Slippery Slope Arguments, *History and Philosophy of Logic*, 14, 1993, 124-128.

63. Review of H.G. Callaway, "Logic Acquisition, Usage and Semantic Realism", *Mathematical Reviews*, 93i, 1993, 4685-4686.

64. Review of F.C. White, On Schopenhauer's Fourfold Root of the Principle of Sufficient Reason, *Canadian Philosophical Reviews*, 12, 1992, 370-372.

65. Review of Paul Gochet and Pascal Gribomont, Logique: Volume 1: méthodes pour l'informatique fondamentale, *The Review of Metaphysics*, 46, 1992, 404-405.

66. Review of Laird Addis, Natural Signs: A Theory of Intentionality, *Canadian Philosophical Reviews*, 11, 1991, 1-3.

67. Review of Edward N. Zalta, Intensional Logic and the Metaphysics of Intentionality, *Philosophy and Phenomenological Research*, 51, 1991, 439-444.

68. Review of Alexius Meinong, Über Gegenstandstheorie/Selbstdarstellung, edited by Josef M. Werle, *Brentano Studien*, 3, 1990-1991, 267-268.

69. Review of Marie-Luise Schubert Kalsi, Meinong's Theory of Knowledge, *Noûs*, 24, 1990, 487-492.

70. Review of Peter Simons, Parts: A Study in Ontology, *Philosophy of Science*, 57, 1990, 540-542.

71. Review of Anthony Palmer, Content and Object: The Unity of the Proposition in Logic and Psychology, *History and Philosophy of Logic*, 10, 1989, 113-115.

72. Review of Roy A. Sorensen, Blindspots, *The Journal of Speculative Philosophy*, 3, 1989, 218-223.

73. Review of Daniel C. Dennett, The Intentional Stance, *Mind*, 97, 1988, 619-624.

74. Review of Roderick M. Chisholm, Brentano and Intrinsic Value, *The Journal of Value Inquiry*, 22, 1988, 331-334.

75. Review of Karel Lambert, Meinong and the Principle of Independence: Its Place in Meinong's Theory of Objects and its Significance in Contemporary Philosophical Logic, *International Studies in Philosophy,* 20, 1988, 92-93.

76. "Henry Thoreau's Stoic Legacy" (Review of Robert D. Richardson, Jr., Henry David Thoreau: A Life of the Mind), *Sierra*, 72, January-February 1987, 150-153.

77. "The Meaning of Animal Rights" (Review of Tom Regan, The Case for Animal Rights), Human Studies: A *Journal of Philosophy and the Social Sciences*, 8, 1985, 389- 392.

7 Miscellaneous and Occasional Writings

1. "Salim Kemal (1948-1999)" (Obituary), *The Journal of Nietzsche Studies*, 17, Spring 1999, 94-95.

2. Editor's Pages, *American Philosophical Quarterly*: "Changing of the Guard" (39, April 2002, 107); "Philosophical Voices" (39, July 2002, 213-214); "Philosophy and Practical Life" (39, October 2002, 303-304); "Trends, Dead Ends, and Going Off the Deep End" (40, January 2003, 1-2); "Philosophical Labels" (40, April 2003, 75-76); "Damn the Torpedoes" (40, October 2003, 249-250); "How to Referee a Philosophy Journal Article" (41, January 2004, 1-3); "The Discreet Charm of Tautologies" (41, April 2004, 85-86); "Working Hypotheses" (41, July 2004, 185-186); "Ways of Loving Wisdom" (41, October 2004, 271-272); "Teaching Philosophy as a Dada Concept" (42, January 2005, 1-3).

3. "Philoscopic Vision", *The Philosophers' Magazine*, 45, 2009, 78.

4. "Jacquette, Dale", *International Directory of Logicians*, edited by Dov Gabbay and John Woods (London: College Publilcations, 2009), 173-175.

5. Stanford University Philosophy Talk radio program on journalistic ethics, featuring Dale Jacquette, Journalistic Ethics: Moral Responsibility in the Media,

live broadcast and media stream discussion with Dale Jacquette, John Perry and Ken Taylor, produced by Ben Manilla; also available for download to mp3 devices, 29 April 2007. See the URL: http://www.philosophytalk.org/pastShows/EthicsinJournalism.html.

6. "Paine, Thomas (1737-1809): Anglo-American Political and Religious Thinker" (http:// enlightenment-revolution.org/index.php/Paine%2C_Thomas), "Voltaire, FranDois- Marie Arouet (1694-1778): French Philosophe" (http://enlightenmentrevolution. org/index.php/ Voltaire%2C_François-Marie_Arouet_de), 18th Century *Online Encyclopedia: Enlightenment and Revolution*, edited by Kevin E. Dodson (The Citadel Publishing, electronic format, 2010).

7. Ontology book quoted and discussed in *Knowledge* (BBC Magazine), 'Does God Exist?' (16-page special) by Robert Matthews, March-April 2011 issue 16, 30-33 (excerpt).

8. "Philosophers Stoned", *The Philosophers' Magazine*, 53, 2011, 46-51.

8 Conference Papers and Invited Talks

1. "Brentano's Ambivalent Empiricisms", Conference on the 100th Anniversary of the Death of Franz Brentano, Graz, Austria, October 2017. [INVITED (not given)]

2. "Metaphilosophical Metaphysics" Conference in Honor of Peter van Inwagen, Warsaw, Poland , 26-28 September 2017. [INVITED (not given)]

3. "Depictions of Evil in Religious Art", European Society for Philosophy of Religion, 21st Biennial Conference on 'Evil', Uppsala, Sweden, 25-28 August 2016.

4. "Santayana on Human Scale in Architecture", The Human in Architecture and Philosophy: Towards an Architectural Anthropology, 3rd International Conference of the International Society for the Philosophy of Architecture (ISPA), Department of Philosophy, University of Bamberg, Bamberg, Germany, 20-23 July 2016.

5. "Aristotle's De Anima Psychology of Noûs Poetikos', World Congress, Aristotle 2400 Years, Aristotle University of Thessaloniki, Interdisciplinary Centre for Aristotle Studies, Thessaloniki, Greece, 23-28 May 2016.

6. "Brentano's Signature Contributions to Scientific Philosophy", Inbegriff — Geneva Seminar for Austro-German Philosophy, Workshop on Austro-German Objects and Scientific Philosophy, University of Geneva, Switzerland, 6 October 2015. [Invited]

7. "Leibniz's Empirical, not Empiricist Methodology", conference on Leibniz - Scientist, Leibniz - Philosopher, University of Wales, Lampeter campus, Trinity Saint David, 3-5 July 2015.

8. "ISTAN-BOOLE: Boole School Istanbul — Boolean Algebraic Logic and Philosophy of Logic", Three-Session workshop, Universal Logic conference (UNILOG'15), Istanbul, Turkey, 21-23 June 2015. [Invited]

9. "Brentano and the Ambiguities of Scientific Philosophy", conference on Philosophy as Science: A Key Idea of the Nineteenth Century, Rijksuniversiteit Utrecht, Utrecht, The Netherlands, 9-11 April 2015. [Invited]

10. "Schopenhauer and the Metaphysics of the Unconscious", Philosophisches Seminar, Philosophische Gesellschaft Zürich, UZH, Zürich, Switzerland, 20 November 2014. [Invited]

11. "Mental Causal Efficacy and Contracausal Freedom of Will in the Psychology and Metaphysics of Moral Responsibility for Action", Conference on The Moral Domain: Conceptual Issues in Moral Psychology, Vilnius, Lithuania, 9-11 October 2014.

12. "Tractatus Objects and the Logic of Color Incompatibility", Workshop on Objects of All Kinds, Lille, France, 22-23 May 2014. [Invited]

13. "Subalternation and Existence Presuppositions in a Simplified Unconventional Square of Opposition", 4th World Congress on the Square of Opposition, Pontifical Lateran University, Vatican, Italy, 5-9 May 2014.

14. "Thoughts on Twin Earth", University of Zürich, Philosophisches Seminar, Lecture Series, 'Gedanken-Experimente. Kann man aus dem Lehnstuhl die Welt erforschen?', Zürich, Switzerland, 19 March 2014. [Invited]

15. "Mentale Verursachung: Gleichzeitig einfacher und komplizierter als jemand denken würde", Institutskolloquium, Universität Bern, Institut für Philosophie, Bern, Switzerland, 13 March 2014. [Invited]

16. "Schopenhauer's Transcendental Idealism in Wittgenstein's Early Philosophy', Hawaii International Conference on Arts & Humanities, 12th Annual Conference, Honolulu, Hawaii, 10-13 January 2014.

17. "Lawlike versus Counterfactual Mental Causation", Universität Saarbrücken, 18 December 2013. [Invited]

18. "Schopenhauer as Metaphysician of Sexual Love", Conference on Schopenhauer, Love, and Compassion, Ghent University, Ghent, Belgium, 17-18 October 2013.

19. "Berkeley's Carriage Argument", International Berkeley Society Conference, The 300th Anniversary of the Publication of Three Dialogues Between Hylas and Philonous, Krakow, Poland, 19-22 August 2013.

20. "Wittgenstein's Therapeutic Anti-Philosophy", World Congress of Philosophy (FISP), Special Round Table Session, British Wittgenstein Society, 'Wittgenstein: Therapy or Post-Therapy?', Athens, Greece, 4-10 August 2013. [Invited]

21. "Schuhmann on Representation in Early Husserl", Colloquium in Honor of Karl Schuhmann, Universiteit Utrecht, Utrecht, NL, 14 June 2013. [Invited]

22. "Semantic Intentionality and Intending to Act", Athens Institute for Education and Research (ATINER), Arts and Sciences Research Division, Philosophy Research Unit, Eighth Annual International Conference on Philosophy, Athens, Greece, 27-30 May 2013.

23. "Margolis on the Progress of Pragmatism", Conference on The Metaphysics of Culture — The Philosophy of Joseph Margolis, Helsinki Collegium of Advanced Studies, Helsinki, Finland, 20-21 May 2013.

24. "Anatomy of a Nonidentity Paradox", Workshop on Identity and Paradox, Lille, France, 11-12 April 2013.

25. "Brentano on Aristotle's Psychology of the Active Intellect", Aristotelische Forschungen im 19. Jahrhundert, Munich, Germany, 28 February - 2 March 2013. [Invited]

26. "Gambrinian Hedonic Synergy: Why Drinking Beer to Accompany Other Pleasures is More Than Just the Sum of its Parts", University of Oporto, Oporto, Portugal, November 2012. [Invited]

27. "Justification and Truth Conditions in the Concept of Knowledge", Conference on Judgment and Justification, University of Tampere, Finland, 24-26 September 2012.

28. "Cantor's Diagonalization and Turing's Cardinality Paradox", Computing, Philosophy and the Question of Bio-Machine Hybrids (AISB/IACAP World Congress), Birmingham, UK, 2-6 July 2012.

29. "Logical Contingency in William of Sherwood's Modal Square of Opposition", Square of Opposition III Conference, Beirut, Lebanon, June 26-29, 2012.

30. "About Nothing", International Colloquium, Objects and Pseudo-Objects: Ontological Deserts and Jungles from Meinong to Carnap, Université de Liège, Liège, Belgium, May 15-16, 2012. [Invited]

31. 'Marx and Industrial Age Aesthetics of Alienation', Conference on Marx and the Aesthetic, University of Amsterdam, NL, May 10-13, 2012.

32. "Picturing Logical Relations in Peirce's Existential Graphs", Mind in Motion & The Body of the Sign: Peirce's Semiotical Pragmatism, Workshop sponsored by the Picture Act and Embodiment Research Group (Kolleg-Forschergruppe, Bildakt und Verkörperung), Humboldt Universität, Berlin, Germany, March 15-17, 2012. [Invited]

33. "Socrates on the Moral Mischief of Misology", Socratica III: A Conference on Socrates, the Socratics, and the Ancient Socratic Literature, Università degli Studi di Trento, International Plato Society, Trento, Italy, February 23-25, 2012.

34. "Practitions in Castañeda's Deontic Logic", Workshop on the Philosophy of Hector-Neri Castañeda, Urbino, Italy, June 23, 2011.

35. "Realism versus Idealism in the Nature-Nurture Dispute', Defending Realism Conference, Urbino, Italy, June 20-23, 2011.

36. "Slingshot Arguments and the Intensionality of Identity", Crossing Borders, Austrian Society for Philosophy, Vienna, Austria, June 2-4, 2011.

37. "Austin on Conceptual Polarity and Sensation Deception Metaphors", J.L. Austin Centenary Conference, Lancaster University, England, April 5-7, 2011.

38. "Philosophical Therapy and Wittgenstein", Symposium on Eugen Fischer, Philosophical Delusion and its Therapy, University of East Anglia, Norwich, England, March 25- 26, 2011. [Invited]

39. "Identity and Fictional Possibilities", Guest Colloquium Lecture, Department of Philosophy, University of East Anglia, Norwich, England, March 24, 2011. [Invited]

40. "Against Epistemic Hypocrisy", Copenhagen-Lund Workshops in Social Epistemology, Copenhagen University, Copenhagen, Denmark, November 25, 2010.

41. "Knowledge Without Truth", Philosophy Colloquium, Institut de philosophie, Faculté des lettres et sciences humaines, Université de Neuchâtel, Switzerland, November 16, 2010.

42. "Belief State Intensity", Conference on The Nature of Belief — The Ontology of Doxastic Attitudes, University of Southern Denmark, Odense, Denmark, October 18- 19, 2010.

43. "Socrates on the Moral Mischief of Misology", Seventh International Conference on Argumentation, International Society for the Study of Argumentation (ISSA), University of Amsterdam, The Netherlands, June 29-July 2, 2010.

44. . "Thinking Outside the Square of Opposition Box", Second World Congress on the Square of Opposition, Corte, Corsica, France, June 17-20, 2010. [Invited]

45. "Dummett on Truth-Makers, Frege's Analysis of Sentence Meaning, and the Slingshot Argument", International Symposium in Honor of Michael Dummett, The Lauener Founda-tion, Bern, Switzerland, May 26-27, 2010. [Invited]

46. "Wittgenstein's Tractatus as Mystic Revelation", Ludwig Wittgenstein Lecture, British Wittgenstein Society (BWS), University of Hertfordshire, Herts, England, May 20, 2010. [Invited]

47. "An Argument for Universal Logic: Relativizing Truth Functions in Tarski's Truth- Schemata Hierarchy", Third Conference on Universal Logic, Lisbon, Portugal, April, 18-25, 2010.

48. "Contemporary and Future Directions of Analytic Philosophy", Panel Commentary, The State and Prospects of Philosophical Research, UNESCO Symposium, organized by Adam Sennet, American Philosophical Association, Pacific Division, San Francisco, CA, March 31 - April 4, 2010. [Invited]

49. "Justification and Truth Conditions in the Concept of Knowledge", Episteme Conference on 'Justification Revisited', Université de Genève, Geneva, Switzerland, March 25- 27, 2010.

50. . "Intentionality as a Conceptually Primitive Relation", Intentionality Workshop, Universität Graz, Graz, Austria, November 20-21, 2009. [Invited]

51. "Causal Relevance in the Metaphysics of Scientific Explanation", 2nd Conference of the European Philosophy of Science Association, Amsterdam, NL, October 21-24, 2009.

52. "Hume's Enlightenment Aesthetics and Philosophy of Mathematics", Short Conference on Hume and the Enlightenment, Adelaide, Australia, July 13-14, 2009.

53. "Causal Relevance in the Metaphysics of Scientific Explanation", Conference on the Meta-physics of Science, University of Melbourne, Australia, July 2-5, 2009.

54. "Self-Justificatory versus Extra-Justificatory Foundations of Knowledge", 19th Inter- University Workshop on Philosophy and Cognitive Science, Zaragoza University, Zaragoza, Spain, May 27-29, 2009.

55. "Anselm's Metaphysics of Nonbeing", International Anselm Conference, University of Kent at Canterbury, England, April 22-25, 2009.

56. "Newton's Noncausal Concept of Emanative Effect", Workshop on Causation, Universiteit Leiden, The Netherlands, April 10-11, 2009.

57. "Ontology of Propositions and Logically Possible Worlds", Conference on Propositions: Ontology, Semantics, and Pragmatics, Venice, Italy, November 17-19, 2008.

58. "Brentano's Scientific Psychology", History of the Philosophy of Science (HOPOS), Vancouver, British Columbia, Canada, June 18-21, 2008.

59. "Toward an Inherentist Philosophy of Mathematical Entities", Joint Paris-Arché Workshop on Abstract Objects in Semantics and the Philosophy of Mathematics, école Normale Supérieure, Paris, France, February 28 - March 1, 2008.

60. "Intensional Truth Functions", Philosophy of Science Center, University of Pittsburgh, Pittsburgh, PA, October 19, 2007.

61. "Intensional Truth Functions", Logic Colloquium, University at Buffalo, State University of New York, Buffalo, NY, September 29 2007.

62. "Formal Criteria of Non-Truth-Functionality", 5th Barcelona Workshop on Issues in the Theory of Reference (BW5), Non-Truth-Conditional Aspects of Meaning, Grup de Recerca en Lògica, Llenguatge i Cognició, University of Barcelona, Barcelona, Spain, June 5-8, 2007.

63. "Burleigh's Paradox", Society for Exact Philosophy, University of British Columbia, Van-couver, BC, Canada, May 17-20, 2007.

64. "Logic and Metaphysics of Negative States of Affairs", Department of Philosophy, Bilkent University, Ankara, Turkey, April 24, 2007. [Invited]

65. "Logic and Metaphysics of Negative States of Affairs", Institut für Philosophie, Philosophisch- historisch Fakultät, Universität Bern, Bern, Switzerland, April 20, 2007. [Invited]

66. "Supervenience (on Steroids) and the Mind", SophiaEuropa, Second Workshop on Causality and Motivation, Pontificio Ateneo San Anselmo, Rome, Italy, April 13-14, 2007.

67. "Logical Structure and Analogical Foundations of Scientific Law", Department of Philosophy, University of Western Ontario, London, Ontario, Canada, March 12, 2007. [Invited]

68. "On the Logic of Negation", Department of Philosophy, University of Aberdeen, Aberdeen, Scotland, February 20, 2007. [Invited]

69. "On the Metaphysics of Negation", Department of Philosophy, Georgia State University, Atlanta, Georgia, January 26, 2007. [Invited]

70. "Deductivism and the Informal Fallacies", Sixth International Conference on Argumentation, International Society for the Study of Argumentation (ISSA), University of Amsterdam, Amsterdam, The Netherlands, June 27-30, 2006.

71. "Intentionality and Intensional Semantics", Philosophy Colloquium, Universität Graz, May 23, 2006. [Invited]

72. Workshop on Kazimierz Twardowski's Phenomenology of Indirect Presentations, Universität Graz, May 22, 2006. [Invited]

73. "Logic and Semantics of False Predications", Philosophy Colloquium, Universität Salzburg, May 18, 2006. [Invited]

74. "Collingwood on Historical Authority and Historical Imagination", Philosophy and Historio-graphy, Robinson College, Cambridge University, England, April 3-5, 2006.

75. "How (Not) to Justify Induction", Argumentation Group Research Colloquium, University of Amsterdam, Amsterdam, The Netherlands, December 2, 2005. [Invited]

76. "Logical Models of Scientific Law", Institute for Logic, Language, and Computing (ILLC), Logic Tea Colloquium, University of Amsterdam, Amsterdam, The Netherlands, Novem-ber 22, 2005. [Invited]

77. "Metaphysics of Mathematics", Zeno Lecture, Joint Philosophy Faculty Colloquium, Leiden and Utrecht Universities, Leiden, The Netherlands, November 4, 2005. [Invited]

78. "Negative States of Affairs", Vakgroep Theoretische Filosofie, Rijksuniversiteit Groningen, The Netherlands, November 2, 2005. [Invited]

79. "Wittgenstein's Tractatus as Mystic Revelation", Philosophy Colloquium, Universität Salz-burg, October 27, 2005. [Invited]

80. "Truth-Breakers", Society for Exact Philosophy, University of Toronto, Toronto, Canada, May 19-22, 2005.

81. "Two Sides of Any Issue", Ontario Society for the Study of Argumentation (OSSA), Mc- Master University, Hamilton, Ontario, Canada, May 18-21, 2005.

82. Commentary on Jan Albert van Laar, "One-Sided Arguments", Ontario Society for the Study of Argumentation (OSSA), Mc-Master University, Hamilton, Ontario, Canada, May 18-21, 2005. [Invited]

83. "Meditations on Meinong's Golden Mountain", Russell versus Meinong: 100 Years After 'On Denoting', The Bertrand Russell Research Centre, McMaster University, Hamilton, Ontario, Canada, May 14-18, 2005. [Invited]

84. "Schopenhauer's Proof that Thing-in-Itself is Will", North American Schopenhauer Society, American Philosophical Association, Central Division, Chicago, IL, April 27-30, 2005.

85. "Analogical Reasoning in the Logical Structure of Scientific Law", Philosophy of Science Center, University of Pittsburgh, Pittsburgh, PA, April 8, 2005.

86. "Applied Mathematics in the Sciences", Philosophy of Science Conference, Thirtieth Annual Meeting, 'Philosophy of Logic and Mathematics', Inter-University Centre, Dubrovnik, Croatia, April 19-24, 2004. [Invited]

87. "Socrates on Rhetoric, Truth, and Courtroom Argumentation in Plato's Apology", Informal Logic at 25, University of Windsor, Windsor, Ontario, Canada, May 14-17, 2003.

88. "Rationality and the Preface Paradox", American Philosophical Association, Association for Informal Logic and Critical Thinking, Pacific Division, San Francisco, CA, March 27-30, 2003.

89. "Nonstandard Modal Semantics and the Concept of a Logically Possible World", International Symposium on Philosophical Insights into Logic and Mathematics, Nancy, France, September 30 - October 4, 2002.

90. "Kripke's Modal Objection to the Description Theory of Reference", Sixteenth International Symposium, LOGICA 2002, Zahradky Castle, Czech Republic, June 18-21, 2002.

91. "Rationality and the Preface Paradox", Conference on Rationality, Bled, Slovenia, June 3-8, 2002. [Invited]

92. "Probability, Practical Reasoning, and Conditional Statements of Intent", Workshop on 'Conditionals, Information, and Inference', Hagen, Germany, May 13-15, 2002.

93. "Realism and Idealism in the Metaphysics of Wittgenstein's Tractatus", Society for Realism/Antirealism, American Philosophical Association, Central Division, Chicago, IL, April 24-27, 2002. [Invited]

94. "Animadversions on the Logic of Fiction and Reform of Modal Logic", Mistakes of Reason: A Conference in Honour of John Woods, The University of Lethbridge, Alberta, Canada, April 19-21, 2002.

95. "Brentano's Scientific Revolution in Philosophy", Spindel Philosophy Conference, 'Origins: The Common Sources of the Analytic and Phenomenological Traditions', University of Memphis, Memphis, TN, September 20-22, 2001. [Invited]

96. "Wittgenstein on Thoughts as Pictures of Facts and the Transcendence of the Metaphysical Subject", Austrian Ludwig Wittgenstein Society, Twenty-Fourth

International Wittgen-stein Symposium, Kirchberg am Wechsel, Austria, August 12- 18, 2001. [Invited Plenary Speaker]

97. "The Importance of Wittgenstein's Tractatus", Panel Discussion on Wittgenstein's Tractatus Logico-Philosophicus, Austrian Ludwig Wittgenstein Society, Twenty- Fourth Interna-tional Wittgenstein Symposium, Kirchberg am Wechsel, Austria, August 12-18, 2001. [Invited Panelist]

98. "Self-Justificatory versus Extra-Justificatory Foundations of Knowledge", The Epistemology of Basic Belief, Vrije Universiteit Amsterdam, Amsterdam, The Netherlands, June 20-22, 2001.

99. "Margolis on History and Nature", Invited Symposium on The Work of Joseph Margolis, American Philosophical Association, Pacific Division, San Francisco, CA, March 28 - April 1, 2001. [Invited]

100. "Fin de Siècle Austrian Philosophy as a Model for International Research", 'Approaching a New Millennium: Lessons From the Past, Prospects for the Future', International Society for the Study of European Ideas (ISSEI), Seventh International Conference, Bergen, Norway, August 14-18, 2000.

101. "Thomas Reid on Natural Signs, Natural Principles, and the Existence of the External World", Second International Reid Symposium, Philosophy in Scotland Then and Now, University of Aberdeen, Scotland, July 10-12, 2000.

102. "Theory and Observation in the Philosophy of Science", Fourth Quadrennial International Fellows Conference on Philosophy and History of Science, University of Pittsburgh Center for the Philosophy of Science, San Carlos de Bariloche, Argentina, June 22- 26, 2000.

103. "David Lewis on Meinongian Logic of Fiction", Writing the Austrian Traditions: A Conference on the Relation Between Austrian Philosophy and Literature, Woodsworth College, University of Toronto, Toronto, Canada, May 12-14, 2000. [Invited]

104. "Psychologism Revisited in Logic, Metaphysics, and Epistemology", Invited Symposium on Psychologism: The Current State of the Debate, American Philosophical Association, Pacific Division, Albuquerque, NM, April 5-8, 2000. [Invited featured speaker]

105. "Of Time and the River in Kant's Refutation of Idealism", IX. Internationaler Kant- Kongress, Berlin, Germany, March 26-31, 2000.

106. "Wittgenstein's Later Conception of the Social Sciences", Philosophy of Science Center, University of Pittsburgh, December 7, 1999.

107. "Identity, Intensionality, and Moore's Paradox", The Creighton Club, New York State Philo-sophical Society, Skaneateles, NY, November 12-13, 1999.

108. "Goodman on the Concept of Style", American Society for Aesthetics, Washington, DC, October 27-30, 1999.

109. "The Soundness Paradox", Eleventh International Congress of Logic, Methodology and Philosophy of Science, Cracow, Poland, August 20-26, 1999.

110. "Schopenhauer on the Ethics of Suicide", North American Schopenhauer Society, American Philosophical Association, Central Division, New Orleans, LA, May 5-8, 1999.

111. Commentary on S.M. Puryear, "A Restriction on Divine Conceptualist Theories", American Philosophical Association, Central Division, New Orleans, LA, May 5-8, 1999. [Invited]

112. "Conundrums of Conditionals in Contraposition", American Philosophical Association, Eastern Division, Washington, DC, December 27-30, 1998.

113. "Carl Stumpf on the Ontology of Relations", Internationale interdisziplinäre Fachkonferenz anläßlich des 150. Geburtstag von Carl Stumpf, Julius Maximilians Universität, Würzburg, Germany, September 30 - October 3, 1998. [Invited]

114. "The Deconstruction Debacle in Theory and Practice", Federation Internationale des Societes de Philosophie, PAIDEIA, Twentieth World Congress of Philosophy, Special Session on 'Deconstruction and its Critics', Boston, MA, August 10-16, 1998. [Invited]

115. "Intentionality on the Installment Plan", Eastern Pennsylvania Philosophical Association, Bloomsburg University, Bloomsburg, PA, October 25, 1997. [Invited Keynote Address]

116. "Constructibility and the Analysis of Quantifiers in Wittgenstein's Tractatus", The Austrian Ludwig Wittgenstein Society, Twentieth International Wittgenstein Symposium, Kirchberg am Wechsel, Austria, August 10-16, 1997.

117. "Paraconsistent Logical Consequence", First World Congress on Paraconsistency, Universiteit Gent, Belgium, July 30 - August 2, 1997.

118. "On the Relation of Informal to Formal Logic", Conference on Argumentation and Rhetoric, Ontario Society for the Study of Argumentation, Brock University, St. Catharine's, Ontario, Canada, May 15-17, 1997.

119. "Ignorance is No Excuse (for Deductively Invalid Inference)", Commentary on Jonathan E. Adler, "Arguing from Ignorance", Conference on Argumentation and Rhetoric, Ontario Society for the Study of Argumentation, Brock University, St. Catharine's, Ontario, Canada, May 15-17, 1997. [Invited]

120. Commentary on Ralph Clark, "Is Time a Series?", American Philosophical Association, Central Division, Pittsburgh, PA, April 23-26, 1997. [Invited]

121. "Sellars on the Intentionality of Thought and Language in Empiricist Philosophy of Mind", Conference on The Philosophy of Wilfrid Sellars: A Seminar on 'Empiricism and the Philosophy of Mind', Dunabogdany, Hungary, 31 October - 3 November 1996. [Invited]

122. "Truth as a Regulative Concept of Philosophical Semantics", Conference on Truth, Inter- University Centre of Dubrovnik, Croatia, Bled, Slovenia, June 3-8, 1996. [Invited]

123. "Intentionality on the Installment Plan", Rijksuniversiteit Utrecht, The Netherlands, May 3, 1996. [Invited]

124. "Wittgenstein on Logic in the Tractatus and 'Some Remarks on Logical Form'", Universidad del Pais Vasco, Institute for Logic, Cognition, Language and Information, Donostia (San Sebastian), Spain, April 23, 1996. [Invited]

125. "Wittgenstein on the Transcendence of Ethics", Universidad del Pais Vasco, Institute for Logic, Cognition, Language and Information, Donostia (San Sebastian), Spain, April 23, 1996. [Invited]

126. . Discussion Panel Participant, "Wittgenstein: Logic and Ethics" (workshops): 'Wittgenstein: Mind and Logic', 'Wittgenstein: Ethics and Philosophy', Universidad del Pais Vasco, Institute for Logic, Cognition, Language and Information, Donostia (San Sebastian), Spain, April 23, 1996. [Invited]

127. "Searle's Antireductionism", Universidad del Pais Vasco, Department of Logic and Philosophy of Science, Donostia (San Sebastian), Spain, April 22, 1996. [Invited]

128. "Neurobiology and the Soul", Ciclo Sobre Ciencia Cognitiva, Universidad Complutense de Madrid, Facultad de Filosofia, Madrid, Spain, April 18, 1996. [Invited]

129. "Meinong's Concept of Implexive Being and Nonbeing", International Meinong Conference 1995, Forschungsstelle und Dokumentationszentrum für österreichische Philosophie, Graz, Austria, September 28-30, 1995. [Invited]

130. Commentary on D. Tycerium Lightner, "Hume on Inconceivability, Impossibility, and Adequate Ideas", Hume Society, Twenty-Second Annual Conference, Park City, Utah, July 25-29, 1995. [Invited]

131. Discussion Panel Participant, David Fate Norton's Critical Edition of David Hume's A Treatise of Human Nature, Hume Society, Twenty-Second Annual Conference, Park City, Utah, July 25-29, 1995. [Invited]

132. "Lloyd on Intrinsic Natural Representation in Simple Mechanical Minds", Society for Exact Philosophy, Silver Jubilee Meeting, The University of Calgary, Calgary, Alberta, Canada, May 25-28, 1995.

133. "Kant on Unconditional Submission to the Suzerain", American Philosophical Association, Central Division, Chicago, IL, April 21-24, 1995.

134. "Private Sensation Particulars and Types in Wittgenstein's Private Language Argument", American Philosophical Association, Eastern Division, Boston, MA, December 27- 30, 1994.

135. "Meinongian Objects", Filozofska Fakulteta, Ljubljana, Slovenia, December 14, 1994. [Invited]

136. "On Defoliating Meinong's Jungle", Meinong and his School, UniversitĹ degli Studi di Trento, Centro Studi per la Filosofia Mitteleuropea, Trento, Italy, December 9-10, 1994. [Invited]

137. "Haller on Wittgenstein and Kant", Austrian Philosophy: International Conference in Honor of Rudolf Haller, University of Arizona, Tucson, AZ, November 17-20, 1994. [Invited]

138. "Virtual Relations", Marvin Farber Conference on the Ontology and Epistemology of Relations, State University of New York (SUNY) at Buffalo, Buffalo, NY, September 16-18, 1994.

139. "Charity and the Reiteration Problem for Enthymemes", Third International Conference on Argumentation, International Society for the Study of Argumentation, University of Amsterdam, Amsterdam, The Netherlands, June 21-24, 1994.

140. "Hume's Phenomenal Atomism in the Inkspot Experiment", The Hume Society, Twenty- First Annual International Conference, Rome, Italy, June 21-24, 1994.

141. "A Meinongian Theory of Definite Description", Society for Exact Philosophy, The University of Texas, Austin, TX, May 12-15, 1994.

142. "The Blue Banana Trick: Dennett on Jackson's Color Scientist", American Philosophical Association, Eastern Division, Atlanta, GA, December 27-30, 1993.

143. "Margolis and the Metaphysics of Culture", American Society for Aesthetics, Santa Barbara, CA, October 27-30, 1993. [Invited]

144. "Wittgenstein's Private Language Argument and Reductivism in the Cognitive Sciences", The Austrian Ludwig Wittgenstein Society, Sixteenth International Wittgenstein Symposium, Kirchberg am Wechsel, Austria, August 15-22, 1993. [Invited]

145. "Reconciling Berkeley's Microscopes in God's Infinite Mind", The British Society for the History of Philosophy and The International Berkeley Society, St. Anne's College, Oxford University, Oxford, England, July 16-18, 1993.

146. "Many-Valued Deontic Predications", Society for Exact Philosophy, York University, Toronto, Canada, May 13-16, 1993.

147. "Many Questions Begs the Question (But Questions Do Not Beg the Question)", American Philosophical Association, Central Division, Chicago, IL, April 21-24, 1993.

148. "On the Designated Student and Related Induction Paradoxes", American Philosophical Association, Pacific Division, San Francisco, CA, March 25-28, 1993.

149. "The Type-Token Distinction in Margolis' Aesthetics", American Society for Aesthetics, Eastern Division, Rhode Island School of Design, Providence, RI, March 19-20, 1993.

150. Commentary on Robert Brandom, "Inferentialism and the Expressive Conception of Logic", American Philosophical Association, Eastern Division, Association for Informal Logic and Critical Thinking, Washington, DC, December 27-30, 1992. [Invited]

151. "Hume on the Aesthetics of Greatness and the Sublime", American Society for Aesthetics, Eastern Division, Philadelphia, PA, October 28-31, 1992.

152. "Hume's Aesthetic Psychology of Distance, Greatness, and the Sublime", The Hume Society, Nineteenth Annual International Conference, Nantes, France, June 29-July 3, 1992.

153. "Kant's Second Antinomy and Hume's Theory of Extensionless Indivisibles", American Philosophical Association, Central Division, Louisville, KY, April 23-26, 1992.

154. Commentary on Kurt Ludwig, "Arguments for the Connection Principle", American Philosophical Association, Pacific Division, Portland, OR, March 25-28, 1992. [Invited]

155. "Who's Afraid of the Turing Test?", Greater Philadelphia Philosophy Consortium, Villanova University, Villanova, PA, October 9, 1991. [Invited]

156. "Democracy and the Perils of Informed Consent", Institute for Advanced Philosophic Research, Conference on the Ethics of Democracy, Estes Park, CO, August 19-23, 1991. [Invited]

157. "The Validity Paradox in Modal S5", Ninth International Congress of Logic, Methodology and Philosophy of Science, Uppsala, Sweden, August 7-14, 1991.

158. "The Validity Paradox in Modal S5", Society for Exact Philosophy, University of Victoria, Victoria, British Columbia, Canada, May 19-21, 1991.

159. "Schopenhauer on the Antipathy of Aesthetic Genius and the Charming", North American Schopenhauer Society, American Philosophical Association, Pacific Division, San Francisco, CA, March 28-30, 1991.

160. Commentary on Lucian Krukowski, "Schopenhauer and Expression", American Society for Aesthetics, The University of Texas, Austin, TX, October 24-27, 1990. [Invited]

161. "Validity, Self-Reference, and the Pseudo-Scotus Pseudo-Paradox", Second International Conference on Argumentation, International Society for the Study

of Argumentation (ISSA), University of Amsterdam, Amsterdam, The Netherlands, June 19-22, 1990.

162. "Self-Reference and Infinite Regress in Philosophical Argumentation" (with Henry W. Johnstone, Jr.), Second International Conference on Argumentation, International Society for the Study of Argumentation (ISSA), University of Amsterdam, Amsterdam, The Netherlands, June 19-22, 1990.

163. "Buridan's Bridge", International Symposium on Zeichen und Zeit, Académie du Midi, Institut für Philosophie, L'Abbaye de Lagrasse, Lagrasse, France, June 3-9, 1990.

164. "Categorical Moral Maxims in Kant's Categorical Imperative", Seventh International Kant Congress, Mainz, Germany, March 28 - April 1, 1990.

165. "Logik, Philosophie der Psychologie, und Meinongs Gegenstandstheorie" (in German), Brentano Forschung, Institut für Philosophie, Universität Würzburg, Würzburg, Germany, January 22, 1990. [Invited]

166. "Gödel Incompleteness in a Tarskian Semantic Hierarchy", Logic Seminar, Marie Curie- Sklodowska University, Lublin, Poland, December 14, 1989. [Invited]

167. "Immanent and Transcendent Intentional Objects in Brentano, Twardowski, and Meinong", Polish Philosophical Society, Lublin, Poland, December 15, 1989. [Invited]

168. "Meinongian Models of Scientific Law", Conference on 'The Theory of Objects in Mitteleurope: The Austrian-Polish Connection: Meinong and Twardowski', Forschungs-stelle für Mitteleuropäische Philosophie, Cracow, Poland, December 12- 14, 1989.

169. "Intentional Relations in Zalta's Intensional Logic", Society for Exact Philosophy, University of Alberta, Edmonton, Alberta, Canada, August 17-19, 1989.

170. "Schopenhauer on the Antipathy of Aesthetic Genius and the Charming", American Society for Aesthetics, Rocky Mountain Division, Santa Fe, NM, July 7-9, 1989.

171. "Demonstratives and the Logic of the Self", International Symposium on Language and Metaphysics, Académie du Midi, Institut für Philosophie, L'Abbaye de Lagrasse, Lagrasse, France, June 5-9, 1989.

172. "Metaphilosophy in Wittgenstein's City", Commemorative Conference on the Centennial of Wittgenstein's Birth, Fairleigh Dickinson College, Rutherford, NJ, April 29, 1989.

173. "The Hidden Logic of Slippery Slope Arguments", American Philosophical Association, Eastern Division, Association for Informal Logic and Critical Thinking, Washington, DC, December 27-30, 1988.

174. . "Moral Value and the Sociobiological Reduction", Fourth International Conference on Social Philosophy, North American Society for Social Philosophy, Somerville College, Oxford University, Oxford, England, August 16-18, 1988.

175. "Moral Value and the Sociobiological Reduction", International Society for Value Inquiry, Arundel, England, August 19-21, 1988.

176. "Schopenhauer's Circle and the Principle of Sufficient Reason", Internationale Schopenhauer- Vereinigung, Hamburg, Germany, May 24-27, 1988.

177. "A Fregean Solution to the Paradox of Analysis", American Philosophical Association, Pacific Division, Portland, OR, March 23-26, 1988.

178. "Ducasse on Knowledge and Aesthetics", American Society for Aesthetics, Pacific Division, Pacific Grove, CA, April 1-3, 1987.

179. "Meinongian Mathematics and Metamathematics", The Austrian Ludwig Wittgenstein Society, Eleventh International Wittgenstein Symposium, Kirchberg am Wechsel, Austria, August 4-13, 1986.

180. "Are Minds Machines?", Nebraska Academy of the Sciences, Lincoln, NE, April 15, 1986. [Invited]

181. "Turnstile Operations in Three-Valued Logic", The Fifteenth International Symposium on Multiple-Valued Logic, Computer Society, Institute of Electrical and Electronics Engineers (IEEE), Technical Committee on Multiple-Valued Logic, Kingston, Ontario, Canada, May 28-30, 1985.

182. "Intentionality and Intentional Connections", Society for Exact Philosophy, University of Georgia, Athens, GA, May 3, 1984.

183. "Margolis on Emergence and Embodiment", American Society for Aesthetics, Eastern Division, State University of New York, College at Buffalo, Buffalo, NY, March 24, 1984.

184. "Hume on the Immortality of the Soul", The Hume Society, Eleventh Annual International Conference, York University, Toronto, Canada, August 24-27, 1982.

185. "Berkeley's Continuity Argument for the Existence of God", Conference on Religious Language and Experience, College of the Holy Cross, Worcester, MA, March 26- 27, 1982.

www.ingramcontent.com/pod-product-compliance
Lightning Source LLC
Chambersburg PA
CBHW080533170426
43195CB00016B/2551